Frontiers in Applied Dynamical Systems: Reviews and Tutorials

Volume 4

Frontiers in Applied Dynamical Systems: Reviews and Tutorials

The Frontiers in Applied Dynamical Systems (FIADS) covers emerging topics and significant developments in the field of applied dynamical systems. It is a collection of invited review articles by leading researchers in dynamical systems, their applications and related areas. Contributions in this series should be seen as a portal for a broad audience of researchers in dynamical systems at all levels and can serve as advanced teaching aids for graduate students. Each contribution provides an informal outline of a specific area, an interesting application, a recent technique, or a "how-to" for analytical methods and for computational algorithms, and a list of key references. All articles will be refereed.

Mason A. Porter • James P. Gleeson

Dynamical Systems on Networks

A Tutorial

Springer

Mason A. Porter
Mathematical Institute and CABDyN
  Complexity Centre
University of Oxford, UK

James P. Gleeson
MACSI
Department of Mathematics and Statistics
University of Limerick, Ireland

ISSN 2364-4532   ISSN 2364-4931   (electronic)
Frontiers in Applied Dynamical Systems: Reviews and Tutorials
ISBN 978-3-319-26640-4   ISBN 978-3-319-26641-1   (eBook)
DOI 10.1007/978-3-319-26641-1

Library of Congress Control Number: 2015956340

Mathematics Subject Classification (2010): 05C82, 37-00, 37A60, 90B10, 90B15, 91D30, 92C42, 60J05, 82C26, 82C41, 82C44, 82C43

Springer Cham Heidelberg New York Dordrecht London

Printed on acid-free paper

Springer International Publishing AG Switzerland is part of Springer Science+Business Media (www.springer.com)

To our teachers, students, postdocs, and collaborators. (We are aware that these are overlapping communities.)

Preface to the Series

The subject of dynamical systems has matured over a period of more than a century. It began with Poincaré's investigation into the motion of the celestial bodies, and he pioneered a new direction by looking at the equations of motion from a qualitative viewpoint. For different motivation, statistical physics was being developed and had led to the idea of ergodic motion. Together, these presaged an area that was to have significant impact on both pure and applied mathematics. This perspective of dynamical systems was refined and developed in the second half of the twentieth century and now provides a commonly accepted way of channeling mathematical ideas into applications. These applications now reach from biology and social behavior to optics and microphysics.

There is still a lot we do not understand and the mathematical area of dynamical systems remains vibrant. This is particularly true as researchers come to grips with spatially distributed systems and those affected by stochastic effects that interact with complex deterministic dynamics. Much of current progress is being driven by questions that come from the applications of dynamical systems. To truly appreciate and engage in this work then requires us to understand more than just the mathematical theory of the subject. But to invest the time it takes to learn a new subarea of applied dynamics without a guide is often impossible. This is especially true if the reach of its novelty extends from new mathematical ideas to the motivating questions and issues of the domain science.

It was from this challenge facing us that the idea for the *Frontiers in Applied Dynamics* was born. Our hope is that through the editions of this series, both new and seasoned dynamicists will be able to get into the applied areas that are defining modern dynamical systems. Each article will expose an area of current interest and excitement and provide a portal for learning and entering the area. Occasionally, we will combine more than one paper in a volume if we see a related audience as we have done in the first few volumes. Any given paper may contain new ideas and results. But more importantly, the papers will provide a survey of recent activity and the necessary background to understand its significance, open questions, and mathematical challenges.

Editors-in-Chief
Christopher K.R.T Jones, Björn Sandstede, Lai-Sang Young

Preface

Origin

Traditionally, much of the study of networks has focused on structural features. Indeed, mathematical subjects such as graph theory have a rich history of investigating network structure, and most early work by physicists, sociologists, and other scholars also focused predominantly on structural features. The beginnings of the field of "network science," which one can characterize as the science of connectivity, also started out by focusing on network structure (i.e., literal connectivity). Although some scholars (e.g., many control theorists) have traditionally stressed the importance of dynamics in their study of networks, many network-science practitioners who were trained in fields like dynamical systems and nonequilibrium statistical mechanics (which are both concerned very deeply with dynamical processes) have written myriad papers that seem to focus predominantly or even exclusively on structure. This is valuable and we ourselves have written papers on network structure, but one also needs to consider dynamics, and it is good to wear a dynamical hat even for investigations whose primary explicit focus is on structure. Indeed, a major purpose for studying network structure is as a necessary prerequisite for attaining a deep understanding of dynamical processes that occur on networks. How do social contacts affect disease and rumor propagation? How does connectivity affect the collective behavior of oscillators? The purpose of our monograph is to provide a tutorial for conducting investigations that explore (and try to answer) those types of questions. We will occasionally discuss network structure in our tutorial, but we are wearing our dynamical-systems hats.

Scope, Purpose, and Intended Audience

The purpose of our monograph is to give a tutorial for studying dynamical systems on networks. We focus on "simple" situations that are analytically tractable, though it is also valuable to examine more complicated situations, and insights from simple scenarios can help guide such investigations. There is a large gap between toy models and real life, and it is crucial to worry about what insights the very simplistic models that we know and love are able to reveal about the much more complicated situations that occur in real life. Our monograph is intended for people who seek to study dynamical systems on networks but who might not have any prior experience with graph theory or networks. We hope that reading our tutorial will convey why it is both interesting and useful to study dynamical systems on networks, how one can go about doing so, the potential pitfalls that can arise in such studies, the current research frontier in the field, and important open problems. We touch on a large number of applications, but we focus explicitly on simple models, rules, and equations rather than on realism or data analysis. We do, however, include pointers to references that consider more realistic scenarios. As the eminent philosopher (and baseball player) Yogi Berra once said, "In theory, there is no difference between theory and practice. In practice, there is."

We expect that our tutorial will be most digestible for people who have already had introductory courses in linear algebra and dynamical systems, and some prior experience with probability will also occasionally be helpful. Despite the many contributions from scholars in fields such as statistical physics and sociology (and others), we do not expect our monograph's readers to have any background whatsoever in such subjects. We hope that our tutorial will provide an entry point for graduate students, sufficiently advanced undergraduate students, postdoctoral scholars, or anybody else from mathematics, physics, or engineering who wants to study dynamical systems on networks. It can also perhaps serve as textbook material for the final parts of a course on dynamical systems or statistical physics. Additionally, our tutorial can also be part of the core material in a course on networks or on appropriate topics within networks (e.g., dynamical systems on networks, to give a "random" example), and ideally experts in dynamical systems and network science will also enjoy and benefit from reading our monograph. We have purposely included numerous pointers to interesting papers to read, and we hope that our tutorial will facilitate readers' ability to critically read and evaluate papers that concern dynamical systems on networks. To give a brief warning, our monograph is *not* a review (or anything close to one) on dynamical systems on networks, and we are citing only a small subset of the existing scholarship in this voluminous area. As a complement to citing "classical" pieces of scholarship in the area, we have also purposely included pointers to very recent papers that discuss ideas that we find interesting. New articles on dynamical systems on networks are published or posted on preprint servers very frequently, so we couldn't possibly cite all of the potentially relevant articles even if we tried. See Chapter 7 for a list of books, review articles, surveys, and tutorials on various related topics.

Oxford, UK   Mason A. Porter
Limerick, Ireland   James P. Gleeson
September 2015

Acknowledgements

We were both supported by the European Commission FET-Proactive project PLEXMATH (Grant No. 317614). MAP also acknowledges a grant (EP/J001759/1) from the EPSRC, and JPG acknowledges funding from Science Foundation Ireland (Grants No. 11/PI/1026, 12/IA/1683, 09/SRC/E1780). We acknowledge the SFI/HEA Irish Centre for High-End Computing (ICHEC) for the provision of computational facilities. We thank Alex Arenas, Vittoria Colizza, Marty Golubitsky, Heather Harrington, Petter Holme, Matt Jackson, Ali Jadbabaie, Vincent Jansen, Zoe Kelly, Heetae Kim, Mikko Kivelä, Barbara Mahler, Dhagash Mehta, Joel Miller, Konstantin Mischaikow, Yamir Moreno, Vladimirs Murevics, Mark Newman, Se Wook Oh, Lou Pecora, Thomas Peron, Sid Redner, Edward Rolls, Stylianos Scarlatos, Ingo Scholtes, Michele Starnini, Bernadette Stolz, Steve Strogatz, and three anonymous referees for helpful comments. Achi Dosanjh and the other editors also gave several helpful suggestions and were very accommodating.

Acknowledgements

Contents

Chapter 1
Introduction: How Does Nontrivial Network Connectivity Affect Dynamical Processes on Networks?

When studying a dynamical process, one is concerned with its behavior as a function of time, space, and its parameters. There are numerous studies that examine how many people are infected by a biological contagion and whether it persists from one season to another, whether and to what extent interacting oscillators synchronize, whether a meme on the internet becomes viral or not, and more. These studies all have something in common: the dynamics are occurring on a set of discrete entities (the *nodes* in a network) that are connected to each other via *edges* in some nontrivial way. This leads to the natural question of how such underlying nontrivial connectivity affects dynamical processes. This is one of the most important questions in network science [228], and it is the core question that we consider in our tutorial.

Traditional studies of continuous dynamical systems are concerned with qualitative methods to study coupled ordinary differential equations (ODEs) [127, 292] and/or partial differential equations (PDEs) [63, 65], and traditional studies of discrete dynamical systems take analogous approaches with maps [127, 292].[1] If the state of each node in a network is governed by its own ODE (or PDE or map), then studying a dynamical process on a network entails examining a (typically large) system of coupled ODEs (or PDEs or maps). The change in state of a node depends not only on its own current state but also on the current states of its neighboring nodes, and a network encodes which nodes interact with each other and how strongly they interact.[2]

---

[1]Of course, nothing is stopping us from placing more complicated dynamical processes—which can be governed by stochastic differential equations, delay differential equations, or something else—on a network.

[2]In addition to current states, one can also incorporate dependencies on some of the previous states or even on entire state histories. As suggested both in this footnote and in the previous one, it is possible to envision scenarios that are seemingly arbitrarily complicated.

© Springer International Publishing Switzerland 2016
M.A. Porter, J.P. Gleeson, *Dynamical Systems on Networks*, Frontiers in Applied Dynamical Systems: Reviews and Tutorials 4, DOI 10.1007/978-3-319-26641-1_1

An area of particular interest (because of tractability and seeming simplicity) is binary-state dynamics on nodes, whose states depend on the states of their neighboring nodes and which often have stochastic update rules. (Dynamical processes with more than two states are obviously also interesting.) Examples include simple models of disease spread, where each node is considered to be in either a healthy (*susceptible*) state or an unhealthy (*infected*) state, and infections are transmitted probabilistically along the edges of a network. One can apply approximation methods, such as mean-field approaches, to obtain (relatively) low-dimensional descriptions of the global behavior of the system—e.g., to predict the expected number of infected people in a network at a given time or as a function of time—and these methods can yield ODE systems that are amenable to analysis via standard approaches from the theory of dynamical systems.

Importantly, it is true not only that network structure can affect dynamical processes on a network, but also that dynamical processes can affect the dynamics of the network itself. For example, when a child gets the flu, he/she might not go to school for a couple of days, and this temporary change in human activity affects which social contacts take place, which can in turn affect the dynamics of disease propagation. We will briefly discuss the interactions of dynamics on networks with dynamics of networks (these are sometimes called "adaptive networks" [124, 272]) in this monograph, but we will mostly assume time-independent network connectivity so that we can focus on the question of how network structure affects dynamical processes that occur on top of a network. Whether this is reasonable for a given situation depends on the relative time scales of the dynamics on the network and the dynamics of the network.

The remainder of our tutorial is organized as follows. Before delving into dynamics, we start by recalling a few basic concepts in Chapter 2. In Chapter 3, we discuss several examples of dynamical systems on networks. In Chapter 4, we give various theoretical considerations for general dynamical systems on networks as well as for several of the systems on which we focus. We overview software implementations in Chapter 5. In Chapter 6, we briefly examine dynamical systems on dynamical (i.e., time-dependent) networks, and we recommend several resources for further reading in Chapter 7. Finally, we conclude and discuss some open problems and current research efforts in Chapter 8.

Chapter 2
A Few Basic Concepts

For simplicity, we frame our discussions in terms of unweighted, undirected networks. When such a network is time-independent, it can be represented using a symmetric adjacency matrix $\mathbf{A} = \mathbf{A}^T$ with elements $A_{ij} = A_{ji}$ that are equal to 1 if nodes i and j are connected (or, more properly, "adjacent") and 0 if they are not. We also assume that $A_{ii} = 0$ for all i, so none of our networks include self-edges.[1] We denote the total number of nodes in a network (i.e., a network's "size") by N. The *degree* k_i of node i is the number of edges that are connected to it. For a large network, it is common to examine the distribution of degrees over all of its nodes. The *degree distribution* P_k is defined as the probability that a node—chosen uniformly at random from the set of all nodes—has degree k, and the *degree sequence* is the set of all node degrees (including multiplicities). The *mean degree* z is the mean number of edges per node and is given by $z = \sum_k k P_k$. For example, classical Erdős–Rényi (ER) random graphs have a Poisson degree distribution, $P_k = \frac{z^k e^{-z}}{k!}$, in the $N \to \infty$ limit.[2] However, many real-world networks have right-skewed (i.e., *heavy-tailed*) degree distributions [55], so the mean degree z only provides minimal information about the structure of a network. The most popular type of heavy-tailed distribution is a *power law* [295], for which $P_k \sim k^{-\gamma}$ as $k \to \infty$ (where the parameter γ is called the "power" or "exponent"). Networks with a power-law degree distribution are often called "scale-free networks" (though such networks can still have scales in them, so the monicker is misleading), and many generative mechanisms—such as de Solla Price's model [68] and the Barabási–Albert (BA) model [16]—produce networks with power-law degree distributions.

---

[1] This is a standard assumption, but it is not always desirable. For example, one may wish to investigate narcissism in people tagging themselves in pictures on Facebook, a set of coupled oscillators can include self-interactions, and so on.

[2] By analogy with statistical physics, the $N \to \infty$ limit is often called a "thermodynamic limit."

© Springer International Publishing Switzerland 2016
M.A. Porter, J.P. Gleeson, *Dynamical Systems on Networks*, Frontiers in Applied Dynamical Systems: Reviews and Tutorials 4, DOI 10.1007/978-3-319-26641-1_2

When studying dynamical processes on networks, it can be very insightful to construct networks using convenient random-graph ensembles (i.e., probability distributions on graphs), including both "realistic" ones and patently unrealistic ones.[3] The effects of network structure on dynamics are often studied using a random-graph ensemble known as the *configuration model* [34, 228]. In this ensemble, one specifies the degree distribution P_k (or the degree sequence), but the network *stubs* (i.e., ends of edges) are then connected to each other uniformly at random. In the limit of infinite network size, one expects a network drawn from a configuration-model ensemble to have vanishingly small *degree–degree correlations* and *local clustering*.[4] It is also important to consider computational implementations (and possible associated biases) of the configuration model and its generalizations [21]. Moreover, note that there exist multiple variants of the configuration model.

Degree–degree correlation measures the (Pearson) correlation between the degrees of nodes at each end of a randomly chosen edge of a network. (The edge is chosen uniformly at random from the set of edges.) Degree–degree correlation can be significant, for example, if high-degree nodes are connected preferentially to other high-degree nodes. This is true in a social network if popular people tend to be friends with other popular people, and one would describe the network as "homophilous" by degree. By contrast, a network for which high-degree nodes are connected preferentially to low-degree nodes is "heterophilous" by degree.

The simplest type of local clustering arises as a result of a preponderance of triangle motifs in a network. (More complicated types of clustering—which need not be local—include motifs with more than three nodes, community structure, and core–periphery structure [64, 228, 259].) Triangles are common, for example, in social networks, so the lack of local clustering in configuration-model networks (in the $N \rightarrow \infty$ limit) is an important respect in which their structure differs significantly from that in most real networks. Investigations of dynamical systems on networks with different types of clustering is a focus of current research [129, 213, 216].

---

[3]Reference [212] gives one illustration of how considering a very unrealistic random-graph ensemble can be crucial for developing understanding of the behavior of a dynamical process on networks.

[4]Strictly speaking, one also needs to ensure appropriate conditions on the moments of P_k as $N \rightarrow \infty$. For example, one could demand that the second moment remains finite as $N \rightarrow \infty$.

Chapter 3
Examples of Dynamical Systems

Myriad dynamical systems have been studied in numerous disciplines and from multiple perspectives, and an increasingly large number of these systems have also been examined on networks.[1] In this chapter, we present examples of some of the most prominent dynamical systems that have been studied on networks. We focus on "simple" situations that are analytically tractable, though studying more complicated systems—typically through direct numerical simulations—is also worthwhile.

Many of the dynamical processes that we consider can of course be studied in much more complicated situations (including on directed networks, weighted networks, temporal networks [141], and multilayer [28, 173] networks), and many interesting new phenomena occur in these situations. In our tutorial, however, we want to keep network structure as simple as possible. We explore ways in which network structure has a nontrivial impact on dynamical processes, but we will only include minimal discussion of the aforementioned complications.[2] When placing a dynamical process on a network, one sometimes refer to that network as a "substrate."

In this chapter, we discuss examples of both discrete-state and continuous-state dynamical systems. For the former, it is important to consider whether to update node states synchronous or asynchronously, so we include an interlude that is devoted to this issue.

---

[1]Some scholars choose to draw a distinction between the terms "dynamical process" (e.g., stochastic processes, percolation processes, etc.) and "dynamical system" (e.g., a coupled set of ordinary differential equations). We purposely do not distinguish carefully between the two terms.

[2]In Chapter 6, we do briefly consider time-dependent network structures, because contemplating the time scales of dynamical processes on networks versus those of the dynamics of the networks themselves is a crucial modeling issue.

© Springer International Publishing Switzerland 2016

M.A. Porter, J.P. Gleeson, *Dynamical Systems on Networks*, Frontiers in Applied Dynamical Systems: Reviews and Tutorials 4, DOI 10.1007/978-3-319-26641-1_3

3.1 Percolation

Percolation theory is the study of qualitative changes in connectivity in systems (especially large ones) as its components are occupied or removed [270, 277]. Percolation transitions provide an archetype for continuous transitions, and there has been a lot of work on percolation problems (especially on lattices but increasingly on more general network structures) from both physical and mathematical perspectives [171, 228]. Many percolation problems are deeply related to models for (both biological and social) contagions [217], such as the susceptible–infected (SI) model for biological contagions (see Sec. 3.2.1) and the Watts threshold model for social contagions (see Sec. 3.3.1).

3.1.1 Site Percolation

The simplest type of percolation problem is *site percolation* (i.e., *node percolation*) [228]. Consider a network, and let each of its nodes be either occupied or unoccupied. One can construe occupied nodes as the operational nodes in a network, whereas unoccupied nodes are nonfunctional. We might select nodes uniformly at random and state that they are unoccupied (i.e., are effectively removed from the network) with uniform, independent probability $q = 1 - p \in [0, 1]$. This is a so-called "random attack" with an *occupation probability* of p (and thus an "attack probability" of $1 - p$). Alternatively, we could be devious and perform some kind of "targeted attack" in which we remove (i.e., set as unoccupied) some fraction of nodes preferentially by degree (which is, by far, the most usual case considered), geodesic node betweenness centrality (a measure of how often a node occurs on short paths), location in the network, or some other network diagnostic. In the limit as the number of nodes $N \to \infty$ in one of these processes, what fraction q_c of the nodes needs to be removed so that the network no longer has a very large connected component—called a *giant connected component (GCC)*—of occupied nodes? A *percolation transition* occurs at the critical occupation probability $p_c = 1 - q_c$ that indicates the point of appearance/disappearance of a GCC, which is defined as a connected network component that scales in linear proportion to N as $N \to \infty$. (Such scaling is called "extensive" in the language of statistical mechanics [277].)

3.1.2 Bond Percolation

In *bond percolation* (i.e., *edge percolation*), one tracks occupied edges instead of occupied nodes. Edges are labeled, independently and uniformly at random, as occupied with probability p, which is called the *bond occupation probability*.

As with site percolation, the primary question of interest is the existence and size of a GCC, where connections occur only using occupied edges. If p is below a critical value p_c, then too few edges are occupied to globally connect the network, and a GCC does not exist. However, above the threshold p_c, there is a GCC, and the number of nodes in the GCC scales in linear proportion to N as $N \to \infty$.

3.1.3 K-Core Percolation

The *K-core* of an unweighted, undirected network is the maximal subset of nodes such that each node is connected to at least K other nodes [275]. It is computationally fast to determine K-cores, and they are insightful for many situations [64, 79]. Every unweighted, undirected network has a *K-core decomposition*. A network's *K-shell* is the set of all nodes that belong to the K-core but not to the $(K + 1)$-core. The network's K-core is given by the union of all c-shells for $c \geq K$, and the network's K-core decomposition is the set of all of its c-shells.

One can examine the K-core of a network as the limit of a dynamical pruning process. Start with a network, and then delete all nodes with fewer than K neighbors. After this pruning, the degree of some of the remaining nodes will have become smaller, so repeat the pruning step to delete nodes that now have fewer than K remaining neighbors. Iterating this process until no further pruning is possible leaves the K-core of the network.

3.1.4 "Explosive" Percolation

A few years ago, Ref. [2] suggested the possibility of an "explosive" percolation process in which the transition (as a function of a parameter analogous to the bond occupation probability p) from a disconnected network to a network with a GCC is very steep and could perhaps even be discontinuous. It has now been demonstrated that the proposed family of "explosive" processes, which are often called *Achlioptas processes*, are in fact continuous (in the thermodynamic limit) [66, 326, 327], but the very steep nature of the transitions that they exhibit has nevertheless fascinated many scholars. One can also generalize these processes to ones that exhibit genuinely discontinuous transitions [84].

Let's consider the simplest type of Achlioptas process. Start with N isolated nodes and add undirected, unweighted edges one at a time. Choose two possible edges uniformly (and independently) at random from the set of $N(N-1)/2$ possible edges between a pair of distinct nodes. One adds only one of these edges, making a choice based on a systematic rule that affects the speed of development of a GCC (as compared to the analogous process in which one picks a single edge, rather than making a choice between two possible edges, using the same original random process). One choice that yields "explosive" percolation is to use the so-called "product rule," in which one always retains the edge that minimizes the product of the sizes of the two components that it merges (with an arbitrary choice when there is a tie).

Investigation of Achlioptas processes using different types of network structures and with different choices of rules—especially with the aim of developing rules that do not use global information about a network—is an active area of research. See [84] for a review of work on explosive percolation.

3.1.5   Other Types of Percolation

There are numerous other types of percolation, and it's worth bringing up a few more of them explicitly. *Bootstrap* percolation is an "infection" process in which nodes become infected if sufficiently many of their neighbors are infected [3, 4, 23, 51]. It is related to the Centola–Macy threshold model for social contagions that we will discuss in Sec. 3.3.1. In *limited path percolation*, one construes "connectivity" as implying that a sufficiently short path still exists after some network components have been removed [195]. To appreciate this idea, imagine trying to navigate a city in which some streets are blocked. The percolation of K-cliques (i.e., completely connected subgraphs of K nodes) has been used to study the algorithmic detection of dense sets of nodes known as "communities" [238]. Various percolation processes have also been studied in several different types of multilayer networks (e.g., multiplex networks and interdependent networks) [28, 173].

3.2   Biological Contagions

One of the standard ways to study biological contagions is through what are traditionally called *compartmental models* (although it would be apt to use such a term to describe a much broader set of models), in which the compartments describe a state (e.g., "susceptible," "infected," or "recovered") and there are parameters that represent transition rates for changing states [37]. The simplest compartmental models are applicable to "well-mixed" populations, in which each individual can meet every other individual and in which each type of state change has a single associated probability. Compartmental models can be described using ODEs if one is considering continuous-time state transitions or using maps if one is considering discrete-time state transitions. One can add complications by incorporating space in the form of diffusion (via a Laplacian operator) or by constructing a *metapopulation model*, in which different populations ("patches") can have different fractions of entities in different states of an epidemic (or have other differences, such as different transition rates between their component states). In a sense, a metapopulation model provides a simple way of incorporating network information, and one can think of each node as representing some subpopulation of the full population rather than as an individual entity. This distinction is important when one wishes to consider a metapopulation model on a network [58] (where, e.g., each node might represent a population and the edges indicate interactions between those populations).

The above frameworks assume that one is examining a well-mixed situation (or, for metapopulation models, a partially well-mixed situation), but it is much more realistic in modern society to consider a network of contacts among agents [18, 86, 99, 202, 220, 228, 234, 242, 312]. To do this, one places a compartmental model on a network, so that each node can be in one of several epidemic states (e.g., "susceptible" or "infected"), and the nodes have update rules that govern how the states change. As we discuss in detail in Sec. 3.5, the updates can occur either *synchronously* or *asynchronously*. In synchronous updating, one considers discrete time, and all nodes are updated at once. By contrast, in asynchronous updating, some small fraction of nodes—often just one node—are randomly chosen for update in each time step dt, or the updating algorithm can be event-driven.

Models of biological contagions are often called "simple contagions" because of the mechanism of an infection passing directly from one entity to another. This is a reasonable toy model of some biological contagions, though of course real life can be significantly more complicated [133]. One can also examine biological contagions on more complicated types of networks (such as temporal [139, 141, 142] and multilayer [173, 271] networks).

3.2.1 Susceptible–Infected (SI) Model

The simplest type of biological epidemic has two states—*susceptible* and *infected*— where healthy nodes are considered "susceptible" (and are in the "S" compartment) because they are not currently infected but can become infected, and "infected" nodes (in the "I" compartment) permanently remain in that state. This yields the *susceptible–infected (SI) model*.

One can define the detailed dynamics of SI models in which several different ways. The most common is to consider a stochastic process in which infection is "transmitted" from an infected node to a susceptible neighbor at a rate λ. The parameter λ is a "hazard rate," and the probability of a transmission event occurring in an infinitesimal time interval dt on a chosen edge that connects an infected node to a susceptible node is $\lambda \, dt$. Suppose that we consider a susceptible node that has m infected neighbors. The probability of this node becoming infected during the time interval dt is then

$$1 - (1 - \lambda \, dt)^m \to \lambda m \, dt \quad \text{as} \quad dt \to 0 \,. \tag{3.1}$$

We therefore say that the *infection rate* for a susceptible node with m infected neighbors is λm.

3.2.2   Susceptible–Infected–Susceptible (SIS) Model

Let's consider a (somewhat more complicated) disease-transmission process in which a node can become susceptible again after becoming infected. (Alas, permanent recovery is still impossible.) This process is known as a *susceptible–infected–susceptible (SIS) model*.

We must now introduce stochastic rules for the recovery of infected nodes (i.e., the transition from the infected state to the susceptible state). This transition is usually modeled as a spontaneous process that is independent of the states of neighbors [242].[3] Consequently, each infected node switches to the susceptible state at a constant rate μ. Therefore, in an infinitesimal time interval dt, the probability for a node to switch from the infected state I to the susceptible state S is $\mu\, dt$.

3.2.3   Susceptible–Infected–Recovered (SIR) Model

Another ubiquitous compartmental model is the susceptible–infected–recovered (SIR) model, in which susceptible nodes can still transition to "going" infected, but infected nodes recover to a state R in which they can no longer be infected [37]. Fatalistic people might let state R stand for "removed" instead of recovered, but we're going to be more positive than that.

As with an SIS process, two rates define the stochastic dynamics. These are the transmission rate λ and the recovery rate μ. They are both defined as we described above for SIS dynamics, but now the recovery process takes nodes from the I state to the R state rather than returning them to the S state. Interestingly, one can relate the steady-state of the basic SIR model on a network to a bond-percolation process [123]. See [132, 170, 306] for additional discussion of the connection between SIR dynamics and bond percolation.

3.2.4   More Complicated Compartmental Models

The contagion models that we have discussed above are interesting to study, and they provide a nice family of tractable examples (including for some analytical calculations, as we will discuss in Chapter 4) to examine the effects of nontrivial network structure on dynamics. They also provide interesting toy situations for biological epidemics, although they are grossly unrealistic for most situations. Nevertheless, they are useful for illustrating several ideas, and in particular they have significant potential value in demonstrating effects of network structure on dynamical processes that can also occur in more complicated epidemic models.

---

[3] As with the SI model on networks, there are many variants of the SIS model and other contagion models, and different variants of the SIS model can have different qualitative characteristics (such as different equilibrium behavior) [224].

One step forward is to do similar investigations of more complicated compartmental models on networks. For example, one can add an "exposed" compartment
to obtain an SEIR model, include age structures, and more [37, 242]. One can
also study metapopulation (and "metacommunity") models on networks [58] to
examine both network connectivity and subpopulations with different characteristics, consider non-Markovian epidemics on networks to explore the effects of
memory [30, 172, 204, 247, 314], and explicitly incorporate the effects on contagion
dynamics of individuals with essential societal roles (e.g., health-care workers)
[273]. One can also examine models in which biological contagions are coupled
with information (e.g., "awareness") spread [100].

Another option is to throw analytics out the window, be data-driven, and conduct
simulations of incredibly detailed and complicated situations while estimating
parameter values (and their uncertainties) from real data (as well as doing direct
data analysis of epidemics). Some of these ideas are explored in Refs. [15, 57, 305].
Ultimately, it is important to make advancements in both simple and realistic
approaches, because they complement each other.

3.2.5 Other Uses of Compartmental Models

A variant set of models involves zombification instead of infections [282], and some
of the particular details of the models are occasionally slightly different to reflect
this different application. Compartmental models have also been used for various
models of social influence and information dissemination [205, 242], though they
are not the most common approach to such topics. (See Sec. 3.3 for a selection of
models that are built specifically to study social influence and related phenomena.)

3.3 Social Contagions

Ideas spread along social networks in a manner that appears to be somewhat
analogous to biological contagions, and the perceived similarity between social
and biological epidemics has led to the adoption of the term "contagion" when
describing social influence and the spread of ideas, innovations, and memes
[50, 154, 155, 266, 308, 311]. It is common to discuss ideas "going viral," and
some empirical studies have suggested that the spread of ideas in a social network
can sometimes be genuinely epidemic-like [330]. Specifically, an epidemic-like
(or "simple") contagion refers to cases in which—much like with a virus or a
disease—exposure to a single source is enough to initiate propagation. Unlike
biological contagions, however, ideas spread in a manner that involves *social
reinforcement*: having 100 friends adopt a behavior (or buy a product, join a
movement, etc.) can be rather different than if only one friend does so. Because
of social reinforcement, social contagions need not just spread discretely (or even
discreetly) from one specific source to another. This is why they are sometimes
called *complex contagions* [50].

Identifying causal mechanisms of the spread of ideas is more difficult than for the spread of diseases, and development and (especially) validation of models is significantly less mature in social contexts than in biological ones. Even discerning whether or not genuine social influence is occurring in a network is extremely challenging [278]. We find it helpful (for, e.g., the development of models) to illustrate this difficulty based on a data stream that one might encounter in real life. Suppose that one starts from the empirical observation that the actors represented by the network nodes are adopting (or at least newly exhibiting) some sort of behavior at different times. For example, suppose that various actors in a network are becoming obese or starting to smoke [53, 54, 197], changing their Facebook profile picture into an equal sign [14, 289, 290], or decorating their Facebook profile picture with a rainbow [206]. It is seemingly common for news outlets to posit such observations as "contagions" (e.g., for the United Kingdom riots of 2011 [83]), although the same observations can result from one or more of the following effects [6, 278]:

(1) Genuine spread via social influence, though this could also be social learning (as what might be spreading in a network could be awareness or knowledge about something, a desire to adopt some behavior, or a combination of them). Nevertheless, there is something that is genuinely spreading in a network.
(2) Homophily: Agents tend to adopt the same behavior because they have some common traits that lead to such a propensity. That is, there is some sort of internal similarity between agents (and that may well even be why some of the network edges exist in the first place), but they happen to be adopting the behavior at different times.
(3) Environment: There is a common external influence on the agents. That is, there is some sort of external similarity (or a covariate) that causes agents in a network to adopt the behavior at different times.

Given observations of agents in a network who adopt some behavior at different times, an important goal is to distinguish the relative importances of the above effects. This is not easy, and control strategies (e.g., legislation) clearly depend on whether the cause of the observations is primarily effect (1), (2), (3), or (most likely) some combination of the three. Naturally, one can also consider a more nuanced classification of mechanisms than the simple one above [7].

To address such issues, it is important to do a lot of data collection (e.g., through surveys, online resources, and other means) along with data analysis and statistics, and the majority of studies of social influence tend to take such a perspective. However, it is *also* important to develop simple, tractable models of social influence—in which the various other phenomena that can lead to observations that resemble those from social influence are, by construction, not confounded with something that genuinely spreads on a network—and to examine such dynamical processes on networks. Such models include a wealth of "simple" dynamical systems on networks. Efforts to construct simple models of social influence date back several decades [20, 69, 98, 122, 310], and they have remained an active topic over the years.

3.3.1 Threshold Models

Let's start by discussing simple threshold models of social influence, which have a percolation-like flavor. (In particular, they resemble bootstrap percolation [4].)

In the 1970s, Granovetter posited a simple threshold model for social influence in a fully mixed population [122] (also see the work of Dozier [81]), and it is natural to consider the effects on network structure on such dynamical processes [165, 169, 310, 311, 328]. A well-known threshold model for social influence on networks is the Watts model [328], which uses a threshold-based rule (which is closely related to Granovetter's rule) for updating the states for nodes on a network. For simplicity, we restrict our discussion of the Watts model and other threshold models to unweighted and undirected networks. However, we note in passing that the Watts model has been generalized to weighted [146], directed [102], temporal [164], and multilayer [333] networks.

In binary-state threshold models (such as the Watts model, the Centola–Macy model [50], and others), each node i has a threshold R_i that is drawn from some distribution and which does not change in time.[4] At any given time, each node can be in one of two states: 0 (inactive, not adopted, not infected, etc.) or 1 (active, adopted, infected, etc.). Although a binary decision process on a network is a gross oversimplification of reality, it can already capture two very important features [237]: interdependence (an agent's behavior depends on the behavior of other agents, as we discuss below) and heterogeneity (differences in behavior are reflected in the distribution of thresholds). Typically, some seed fraction $\rho(0)$ of nodes is assigned to the active state, although that is not always true (e.g., when $R_i < 0$ for some nodes i). Depending on the problem under study, one can choose the initially active nodes via some random process (typically, uniformly at random) or with complete malice and forethought. For the latter, for example, one can imagine planting a rumor at specific nodes in a network.

The states of the nodes change in time according to an update rule. As with the models of biological contagions in Sec. 3.2, one can update nodes either synchronously or asynchronously. The latter, which leads naturally to approximations in terms of continuous-time dynamical systems, will be our main focus. When updating the state of a node in the Watts model, one compares the node's fraction m_i/k_i of infected neighbors (where m_i is the number of infected neighbors and k_i is the degree of node i) to the node's threshold R_i. If node i is inactive, it then becomes active (i.e., it switches to state 1) if $m_i/k_i \geq R_i$. Otherwise, its state remains unchanged.

A similar model to the Watts model is the Centola–Macy model [50], in which one considers a node's total number m_i of active neighbors rather than the fraction of such neighbors. (One then writes $m_i \geq R_i$ and makes other similar changes to the formulas that we wrote above.) The special case in which $R_i = R$ for all i in the Centola–Macy model is equivalent to bootstrap percolation [4].

---

[4]Given that people change, it is also relevant to consider models with time-dependent thresholds.

Both the Watts model and Centola–Macy model have a monotonicity property: once a node becomes infected, it remains infected forever. As we will discuss below, this feature is particularly helpful when deriving accurate approximations for the global behavior of these models. The Centola–Macy update rule makes it easier than the Watts update rule for hubs to become active, and this can lead to some qualitative differences in dynamics. It is useful to ponder which of these two toy models provides a better caricature for different applications. For posts on Facebook, for example, one can speculate that the number of posts about a topic might make more of a difference than the fraction of one's Facebook friends that have posted about that topic.

Scholars have also studied several more complicated threshold models. For example, one recent interesting threshold model [207] decomposed the motivation for a node to adopt some behavior as a weighted linear combination of three terms: (1) personal preference, (2) an average of the states of its neighbors, and (3) a system average, which is a measure of the current social trend. It is also interesting to incorporate "synergistic" effects from nearby neighbors into update rules [250]. Another recent study [212] allowed nodes to be in one of three states: 0 (inactive), 1 (active), and 2 (hyper-active).[5] In this so-called "multi-stage" complex contagion, each node has two thresholds. An inactive node exerts no influence, an active node exerts some influence, and a hyper-active node exerts both regular influence and some bonus influence. (A hyper-active node is necessarily active, so state 2 is a subset of state 1. However, state 1 is disjoint from state 0.) In such a multi-stage generalization of the Watts model, a node updates its state when its "peer pressure" $P = [l_1 + \beta l_2]/k$ equals or exceeds a threshold. (The multi-stage version of the Centola–Macy model has a peer pressure of $P = l_1 + \beta l_2$.) The number of neighbors in state i is l_i, β is the bonus influence. In [212], a node whose peer pressure equals or exceeds the first threshold $R_j^{(1)}$ achieves state 1, and a node whose peer pressure equals or exceeds the second threshold $R_j^{(2)} \geq R_j^{(1)}$ achieves state 2. Note that this multi-stage complex contagion model is still monotonic. Some studies have considered non-monotonic generalizations of the Watts and similar threshold models—e.g., by including "hipster" nodes that become inactive if too many of their neighbors are active [76]. It is also interesting to examine the effects of incorporating "Luddite" nodes [211], which can help prevent infections in parts of a network.

3.3.2 Other Models

Although the threshold models of social influence that we described above have the advantage of being mathematically tractable (at least when suitable approximations hold, as we discuss in Chapter 4) and providing a nice (and convenient) caricature of adoption behavior, they are exceptionally simplistic. For example, these threshold

---

[5]Reference [176] also examined a multi-stage update mechanism, although it does not include network structure.

models do not consider specific signals from nodes (e.g., an individual tweeting about the same thing multiple times), which is a "more microscopic" aspect of human behavior. Social reinforcement can arise from multiple friends adopting the same or similar behavior, but it can also arise from the same person sending multiple signals. Moreover, the threshold models are "passive" in the sense that the only signal that ever arises from a node is whether or not it exhibits a behavior (i.e., what state it is in), and the dynamics focus entirely on whether a node is sufficiently influenced (purely from what state other nodes are in) to adopt a new behavior. Reference [254] tried to address this situation by adopting an idea from neuroscience [91] by supposing that each person is an integrate-and-fire oscillator, where the "firing" corresponds to actively sending a signal (e.g., sending a tweet). This can then lead to other nodes adopting the behavior and sending their own signals. Recent models of tweeting have examined the spread of ideas and memes, with an emphasis on the competition between memes for the limited resource of user attention [111, 113, 329].

Approaches other than threshold models to modeling social influence and social learning also date back at least to the 1970s, and some of the models that have been studied are also analytically tractable (although they have a rather different flavor from threshold models). For example, in the DeGroot model [69], individual j has opinion y_j, and the discrete-time opinion dynamics satisfy the equation

$$\mathbf{y}(t+1) = \mathbf{W}\mathbf{y}(t), \qquad t = 0, 1, 2, \ldots, \tag{3.2}$$

where $\mathbf{W}$ is a row-stochastic weight matrix (so that $\sum_j w_{ij} = 1$ for all i), and the matrix element w_{ij} (including the case $i = j$) represents the influence of node j on node i. Friedkin and coauthors (and others) have generalized the DeGroot model in many ways [98, 161].

There are, of course, many other models (see, e.g., the discussion in the introduction of Ref. [212]), so our treatment should not be viewed as even remotely exhaustive. It is interesting to study interactions between biological and social contagions [100] and to consider non-Markovian models of social contagions [324]. There have also been efforts to develop models that attempt to unify biological and social contagions [78], and the percolation-like flavor of threshold models naturally makes it desirable to compare them directly to percolation processes [217]. See Refs. [35, 155, 242] for additional discussions of social contagions.

3.4 Voter Models

Another well-known family of dynamical systems that are often studied on networks are so-called *voter models* [48]. Voter dynamics were first considered by Clifford and Sudbury [56] in the 1970s as a model for species competition, and the dynamical system that they introduced was dubbed the "voter model" by Holley and Liggett a couple of years later [137]. Voter dynamics are fun and versatile (and are very interesting to study on networks), though it is important to ask whether one can ever genuinely construe the voter model (or its variants) as a model for voters [94].

The standard (or "direct") voter model is defined on a network as follows. Each node is associated with a binary variable that can either be in state $+1$ or in state -1. (For example, the former might represent the US Democratic party, and the latter might represent the US Republican party.) At every discrete time step, one node (say, node i) is selected uniformly at random, and node i then adopts the opinion s_j of one of its neighbors j (which is selected uniformly at random from among all of i's neighbors). If i and j were already voting in the same way before the time step, then no change occurs. One can map the standard voter model to a model of random walkers that coalesce when they encounter each other [56, 137]. An alternative to the direct voter model is the "edge-update" voter model [296], in which one chooses an edge (rather than a node) uniformly at random at each time step. If the opinions of the nodes at the two ends of the chosen edge are different, then one randomly selects one of the nodes, and that node adopts the opinion of the other node. The standard voter model and the edge-update voter model have different conserved quantities [287], so we expect them to behave differently from each other. As discussed in Ref. [48], there are a wealth of studies (including mathematically rigorous ones) on voter models.

The original voter model is of course a gross oversimplification of reality, but it is analytically tractable and provides a foundation for numerous interesting generalizations. Indeed, there are a large number of variants of the original voter model, and many of them provide fodder for wonderfully snarky jokes (e.g., see below), and this is especially true if one chooses to label the opinions of nodes with terminology such as "infected" (as is sometimes tempting in political discussions). These models also grossly oversimplify reality, but they are fascinating, are sometimes mathematically tractable (depending on the network structure under consideration), and can even yield insights that are legitimately interesting for applications. For example, whether consensus is reached and how long it takes to reach consensus (or other equilibrium states) depends both on the specific dynamics and on the network on which those dynamics occur. For the direct voter model on configuration-model networks with a power-law degree distribution, the mean consensus time scales linearly with the number N of nodes in the network if the exponent γ of the degree distribution exceeds 3, whereas it scales sublinearly with N if $\gamma \leq 3$ [286, 287]. By contrast, the edge-update voter dynamics can have different asymptotic properties. For example, in a BA network (which has a power-law degree distribution as $N \to \infty$), the consensus time depends linearly on N for any exponent [47]. See [152] for a discussion of consensus times for the direct and edge-update voter models and of which network structures maximize those times.

One nice variant voter model is a "constrained" voter model [317] (see also the more general "political positions process" in [151] and recent work such as [183]), in which nodes can be in one of three states (Left, Right, and Center). All interactions in the constrained voter model involve centrists, as extremists refuse to talk to each other. By considering this model on a complete graph and thereby examining the mean-field limit (see Sec. 4.3 for a discussion of mean-field and related approximations) in which the voters are perfectly mixed, Vázquez et al. [316] derived probabilities, which depend on the initial conditions, of reaching a consensus in one of the three states or of achieving a mixture of the two extremist states. Another interesting variant of the voter model is the (two-state) "vacillating" voter model [180], in which the node i that has been selected examines the states of two of its neighbors, and it changes its state if either of them is different from its own state.[6] See Ref. [48] for discussions of many more types of voter models. Additionally, considering voting and opinion dynamics on multilayer networks allows the formulation of models with especially rich dynamics. (See [22] for an illustration.) Examining voter models (and opinion and influence models) on networks with community structure has the potential to help provide insights into the so-called "majority illusion" [187].

3.5 Interlude: Asynchronous Versus Synchronous Updating

Before discussing additional types of dynamical processes on networks, it is useful to pause and examine how to implement update rules in discrete-state dynamics.

When simulating (stochastic or deterministic) discrete-state dynamics on a network, it is necessary to select a method for choosing which nodes to update and when to update them. Some dynamical processes are defined in a way that is simple to simulate numerically. The voter model described in Sec. 3.4, for example, is defined explicitly in terms of discrete time steps, and one node is chosen uniformly at random to update in each time step. This is a form of *asynchronous updating*, whose monicker reflects the fact that individual nodes are updated independently, so that the new state of a node becomes visible to its neighbors before they attempt to update their own states. One can also employ asynchronous updating for the Watts threshold model described in Sec. 3.3.1. One again chooses a node uniformly at random in each time step and—if it is in the inactive state—one compares the fraction of its active neighbors to its threshold to determine if it becomes active. Alternatively, one can choose to update the states of all nodes simultaneously in each time step; this is called *synchronous* (or *parallel*) updating. When updating in this way, nodes change their states based on the states of their neighbors from the previous time step. If updating nodes using discrete time steps (as is common in computer simulations), one can construe the above synchronous and asynchronous schemes as limiting cases of a more general update scheme in which a fraction f of the nodes are chosen uniformly at random in each time step and one then updates these particular nodes synchronously. If $f = 1/N$, then (on average) one node is updated in each time step, giving the asynchronous methods that we described above (as used, for example, in voter models). The choice $f = 1$ gives synchronous updating.

Synchronous updating has the advantage of allowing fast simulations. By contrast, asynchronous updating admits gradual changes, because only one node is updated per time step, so the fraction of nodes that change state is at most $1/N$ in

---

[6]As an example of a snarky joke, one might imagine that this model is more realistic in some countries than in others. Identification of any such countries is left as an exercise for the diligent reader.

a time step. Consequently, asynchronous updating can lead—in the limit in which a vanishingly small fraction of the nodes are updated in each discrete time step—to dynamics that can be described accurately using a continuum approximation (e.g., by a set of coupled differential equations). For certain classes of dynamics, such as the monotonic dynamics of Sec. 3.3.1, the steady-state (i.e., $t \to \infty$) limits obtained using either synchronous or asynchronous updating schemes are identical, but this need not be true in general. Moreover, the finite-time dynamics are clearly different for asynchronous versus synchronous updating even when the $t \to \infty$ limits are identical. See also [95], which discusses limitations of discrete-time approaches to continuous-time contagion dynamics.

For stochastic dynamical processes (such as the biological contagion models of Sec. 3.2), which are defined in terms of hazard rates, some care is needed in the implementation of an update rule in computational simulations [138]. If a given node i has a rate F_i (i.e., a hazard rate) for changing states, then it has a probability of $F_i \, dt$ of changing its state during an infinitesimal time interval dt. The "infinitesimal" part of this definition is important: it requires that the discrete time step dt of simulations is very small. In practice, dt should be sufficiently small so that only one (or at most a few) nodes are updated in each step. Implementing an update rule in this way ensures that the underlying processes are reproduced faithfully. For example, the time length T that a node spends in its current state before being updated (assuming that no neighbors are updated during this time) should be exponentially distributed. To see that this is reproduced in simulations, note that for each time step of length dt, the probability of node i not changing its state is $1 - F_i \, dt$. Because there are T/dt discrete time steps in the interval $[0, T]$, the probability that the node survives until time T without changing state is the product of the survival probabilities in each step:

$$\text{Prob(survival until at least } T) = (1 - F_i \, dt)^{\frac{T}{dt}} . \qquad (3.3)$$

In the $dt \to 0$ limit, this yields the exponential distribution of survival times (as expected for a Markov process,[7] where the probability of changing state depends only on the current state of the system [30, 314]):

$$\lim_{dt \to 0} (1 - F_i \, dt)^{\frac{T}{dt}} = \exp(-F_i T) . \qquad (3.4)$$

One can use the interpretation of stochastic transition rates in terms of survival times to consider alternative asynchronous updating methods, such as Gillespie or Kinetic Monte Carlo algorithms [30, 105], which are event-driven rather than using equally-spaced time steps. Although they are not entirely straightforward to code, Gillespie algorithms can considerably accelerate simulation times for certain dynamical processes, and we expect their use for stochastic dynamics on networks (including for non-Markovian dynamics [30, 204] and dynamical processes on temporal networks [319]) to become increasingly popular.

---

[7]Naturally, it is also important to consider the effects of memory on dynamical processes on networks [172, 182].

3.6   Coupled Oscillators

Coupled oscillators are a heavily studied type of dynamical system, and associating each oscillator with a node of a network allows one to investigate how nontrivial connectivity affects collective phenomena such as synchronization [10]. Perhaps the most famous model of coupled oscillators is the *Kuramoto model* [1, 10, 128, 178, 265, 293] of phase oscillators. It is one of the canonical models to use in the study of *synchronization* [255, 256], which refers to an adjustment of rhythms of oscillating objects due to their (possibly weak) interactions with each other. The Kuramoto model is also one of the most popular dynamical systems to study on networks. Because each node is associated with an oscillator, it is rather different from the contagion and voter models that we discussed previously.

In the Kuramoto model, each node i has an associated phase $\theta_i(t) \in [0, 2\pi)$ whose dynamics are governed by

$$\dot{\theta}_i := \frac{d\theta_i}{dt} = \omega_i + \sum_{j=1}^{N} b_{ij} A_{ij} f_{ij}(\theta_j - \theta_i), \qquad i \in \{1, \ldots, N\}, \qquad (3.5)$$

where the natural frequency ω_i of node i is typically drawn from some distribution $g(\omega)$ (though it can also be deterministic), $\mathbf{A} = [A_{ij}]$ is the adjacency matrix of an unweighted network, b_{ij} gives the coupling strength between oscillators i and j (so that $b_{ij} A_{ij}$ gives an element of an adjacency matrix of a weighted network), and $f_{ij}(y)$ is some coupling function that depends only on the phase difference between oscillators i and j.

Equation (3.5) is much more general than the traditional Kuramoto model, for which $f_{ij}(y)$ is the same function $f(y)$ for all node pairs, the coupling function is $f(y) = \sin(y)$, and $b_{ij} = b$ for all node pairs. The traditional networks on which to study the Kuramoto model have either all-to-all coupling or nearest-neighbor coupling, but it is both very popular and very interesting to examine the Kuramoto model on networks with more general architectures [1, 10, 128, 159, 265]. The properties of $g(\omega)$ have a significant effect on the dynamics of Eq. (3.5). For example, it is important whether or not $g(\omega)$ has compact support, whether or not it is symmetric, and whether or not it is unimodal. In traditional studies of the Kuramoto model, $g(\omega)$ is unimodal and symmetric about some mean frequency Ω. The original Kuramoto model also uses all-to-all coupling.

Let's briefly consider the original Kuramoto model. To study it, one can track deviations from the mean oscillator frequency Ω by transforming to a rotating frame. (The quantity $v_i := \omega_i - \Omega$ gives the deviation of oscillator i's natural frequency from the mean frequency.) One can thereby see directly which nodes are oscillating faster than the mean frequency and which ones are oscillating slower than the mean frequency. One then defines a complex "order parameter"[8] [10, 293]

---

[8]In statistical physics, an *order parameter* is a quantity (e.g., a scalar) that summarizes a system and is used to help identify and measure some kind of order [277].

$$r(t)e^{i\psi(t)} := \frac{1}{N}\sum_{j=1}^{N} e^{i\theta_j(t)} , \qquad (3.6)$$

where $r(t) \in [0, 1]$ measures the coherence of the set of oscillators and $\psi(t)$ gives their mean phase. The quantity $r(t)$ quantifies the extent to which the oscillators exhibit *phase-locking*, which is a form of synchrony in which (as the name implies) the phase differences between each pair of oscillators have the same constant value. The oscillators are phase-locked when $r(t) = 1$, and they are completely incoherent when $r(t) = 0$. However, these extreme situations only occur in the thermodynamic ($N \to \infty$) limit. In practice, $r(t) \approx 1$ (rather than $r(t) = 1$) when one is considering a finite number of synchronized oscillators, and $r(t) \approx 0$ for a finite number of completely incoherent oscillators. The finite-size fluctuations have a size of $O(1/\sqrt{N})$ [293]. Additionally, one can affect synchronization properties in interesting ways by perturbing the Kuramoto model with noise (which can either promote or inhibit synchrony, depending on the precise details) [179].

When placing Kuramoto oscillators on a network, one can then ask the usual question: how does nontrivial network topology (i.e., connectivity) affect the synchronization dynamics of the oscillators [10, 128]? In addition to numerical simulations, one can conduct analytical investigations using generalizations of the order parameter in (3.6) along with concomitant calculations (see, e.g., [147, 148, 184, 263]) that are more intricate versions of what has been used in studies of the original Kuramoto model [10, 178, 293]. Ideas from spectral graph theory and control theory are also helpful for examining the stability of synchronous dynamics in the Kuramoto model on networks [159].

A particularly interesting phenomenon that can occur in coupled Kuramoto oscillators on networks is *explosive synchronization* [121], which was motivated by prior studies of explosive percolation (see the discussion in Sec. 3.1.4) [2]. Reference [121] elucidated a situation that can lead to a genuinely "explosive" (i.e., discontinuous or "first-order") phase transition in a set of interacting Kuramoto oscillators,

$$\dot{\theta}_i = \omega_i + b \sum_{j=1}^{N} A_{ij} \sin(\theta_j - \theta_i) , \qquad i \in \{1,\ldots,N\} , \qquad (3.7)$$

on a family of networks (see Ref. [120] for a precise specification) that interpolates between BA networks in one limit and ER networks in the other limit. This family of networks is parametrized by one parameter, which we denote by α. One obtains a BA network when $\alpha = 0$ and an ER network when $\alpha = 1$. Suppose that the oscillator frequency $\omega_i \propto k_i^\beta$ (where $\beta > 0$) is positively correlated with node degree. In contrast to most studies of Kuramoto oscillators, these natural frequencies are deterministic rather than chosen randomly from a (nontrivial) distribution. Plotting $r(t)$ from Eq. (3.6) versus the coupling strength b illustrates a phase transition that appears to become discontinuous in the $\alpha \to 0$ limit. The positive correlation between the node degrees and the natural frequencies of the oscillators seems to lead to a positive feedback mechanism that results in a discontinuous phase transition.

To verify that one can truly obtain a discontinuous phase transition (and hence a genuinely "explosive" synchronization transition[9]), let's consider a star network. In such a network, there is a central hub node that is adjacent to all other nodes (the "leaves"), which are each adjacent only to the hub. Suppose that there are $N = K+1$ nodes, so that the hub has degree K and the K leaf nodes each have degree 1. Denote the natural frequency of the hub oscillator by ω_h, and let each leaf node have a natural frequency of ω.

We let $\varphi(t) = \varphi(0) + \Omega t$, where

$$\Omega = \frac{K\omega + \omega_h}{K + 1} \tag{3.8}$$

is (as usual) the mean frequency of the oscillators. We take $\varphi(0) = 0$ without loss of generality, because we can uniformly shift the phases of all oscillators. We thus transform the angular variables as follows:

$$\varphi_h := \theta_h - \Omega t, \qquad j = K+1 \text{ (hub node)},$$
$$\varphi_j := \theta_j - \Omega t, \qquad j \in \{1, \ldots, K\}, \tag{3.9}$$

where we have labeled the $(K+1)$th node using "h" because it is the hub. The equations of motion in Eq. (3.7) thus become

$$\frac{d\varphi_h}{dt} = (\omega_h - \Omega) + b \sum_{j=1}^{K} \sin(\varphi_j - \varphi_h),$$

$$\frac{d\varphi_j}{dt} = (\omega - \Omega) + b \sin(\varphi_h - \varphi_j), \qquad j \in \{1, \ldots, K\}. \tag{3.10}$$

As usual, we define the order parameter using Eq. (3.6) and hence write

$$r(t)e^{i\psi(t)} := \frac{1}{K+1} \sum_{j=1}^{K+1} e^{i\varphi_j(t)} \equiv \langle e^{i\varphi} \rangle, \tag{3.11}$$

where we note that one can express the complex order parameter in terms of an ensemble average over the oscillators. In our current coordinates (a rotating reference frame), the mean oscillator phase is 0, so $\psi = \varphi(0) = 0$, and separately equating real and imaginary parts in Eq. (3.11) yields

---

[9]Recall from Sec. 3.1.4 that the "explosive" percolation process that we described is actually a steep but continuous transition [66, 84, 326, 327]. Fizz.

$$r(t) = \frac{1}{K+1} \sum_{j=1}^{K+1} \cos \varphi_j(t) \equiv \langle \cos \varphi(t) \rangle ,$$

$$0 = \frac{1}{K+1} \sum_{j=1}^{K+1} \sin \varphi_j(t) \equiv \langle \sin \varphi(t) \rangle . \tag{3.12}$$

We multiply Eq. (3.11) by $e^{-i\varphi_h(t)}$ to obtain

$$r e^{i(\psi - \varphi_h)} = r e^{-i\varphi_h} = \frac{1}{K+1} \sum_{j=1}^{K+1} e^{i(\varphi_j - \varphi_h)} , \tag{3.13}$$

which implies that

$$r \cos \varphi_h - i r \sin \varphi_h = \frac{1}{K+1} \sum_{j=1}^{K+1} \cos(\varphi_j - \varphi_h) + i \frac{1}{K+1} \sum_{j=1}^{K+1} \sin(\varphi_j - \varphi_h)$$

$$= \frac{1}{K+1} \left[1 + \sum_{j=1}^{K} \cos(\varphi_j - \varphi_h) \right] + i \frac{1}{K+1} \sum_{j=1}^{K} \sin(\varphi_j - \varphi_h) , \tag{3.14}$$

where we have separated the hub term (with $j = K + 1 = h$) from the other terms in the sums. (Note that we are now suppressing the explicit indication of time-dependence for quantities such as $r(t)$ and $\varphi(t)$.) We separately equate the real and imaginary parts of Eq. (3.14), and the latter yields

$$-r \sin \varphi_h = \frac{1}{K+1} \sum_{j=1}^{K} \sin(\varphi_j - \varphi_h) , \tag{3.15}$$

which we insert into the first equation in Eq. (3.10) to obtain[10]

$$\frac{d\varphi_h}{dt} = (\omega_h - \Omega) - b(K + 1) r \sin(\varphi_h) , \tag{3.16}$$

where the second term is a mean-field coupling term because all interactions with other oscillators depend only on the ensemble average. As we are using a mean-field approach, we are thinking of $K > 0$ as large.

---

[10]Note that the analog of Eq. (3.16) in Ref. [121] has a sign error.

The hub oscillator is phase-locked when $\frac{d\varphi_h}{dt} = 0$ (i.e., when the relative phases are constant), which occurs when

$$\sin(\varphi_h) = \frac{\omega_h - \Omega}{b(K+1)r}.$$ (3.17)

For leaf oscillators to be phase-locked, we need $\dot{\varphi}_j = 0$ for all $j \in \{1, \ldots, K\}$. We thus require that

$$\cos \varphi_j = \frac{1}{b} \left\{ (\Omega - \omega) \sin \varphi_h \pm \sqrt{\left[1 - \sin^2(\varphi_h)\right]\left[b^2 - (\Omega - \omega)^2\right]} \right\}, \quad j \in \{1, \ldots, K\},$$

which is valid as long as $\Omega - \omega \leq b$. We lose phase-locking at a critical coupling of $b = b_c := \Omega - \omega$. For example, if $\omega_h = K$ and $\omega = 1$ (i.e., when each oscillator frequency is equal to the degree of the associated node), we obtain

$$\Omega = \frac{2K}{K+1}, \qquad b_c = \frac{K-1}{K+1}.$$ (3.18)

The critical value r_c of the order parameter r is then[11]

$$r_c = r(b = b_c) = \left. \frac{\cos \varphi_h + K \cos \varphi_j}{K+1} \right|_{b=b_c} = \frac{K}{K+1} > 0.$$ (3.21)

Because $r_c > 0$, we obtain a vertical gap (i.e., a discontinuous synchronization transition) in the phase diagram (i.e., bifurcation diagram) of r versus b. When $K \to \infty$, the critical value $r_c \to 1$, so there is a discontinuous (i.e., "explosive") phase transition from no synchrony (i.e., complete incoherence) to complete synchrony in the thermodynamic limit. Kaboom!

---

[11]Using an approach based on phase-locking manifolds, Zou et al. [334] calculated that

$$r^2 = \frac{K^2 + 1}{(K+1)^2} + \frac{2K}{(K+1)^2} \sqrt{1 - \left(\frac{(K-1)}{b(K+1)}\right)^2},$$ (3.19)

from which they determined that

$$r_c = \frac{\sqrt{K^2 + 1}}{K+1}$$ (3.20)

is the value of r for which the phase-locking manifold (and hence the phase-locking solution) ceases to exist. The expressions for r_c in Eqs. (3.21) and (3.20) approach each other as $K \to \infty$, so they give the same result for explosiveness. However, as $K \to 0$, the expression in Eq. (3.21) approaches $r_c = 0$ instead of giving the correct limiting value of $r_c = 1$. (Equation (3.21) was derived using a mean-field argument, and $K \to 0$ leaves only the hub, which is very far away from a mean-field setting.) The expression in Eq. (3.20) correctly gives $r_c \to 1$ as $K \to 0$.

3.7 Other Dynamical Processes and Phenomena

Numerous other dynamical systems have also been studied on networks, and obviously many others can be. In this section, we briefly discuss a few of these other dynamical processes. As with most of our other discussions, we are not being even remotely exhaustive.

General ideas for dynamical systems on networks (see Sec. 4.1 for an example of one type of methodology) have been used to examine stability for a wealth of both continuous and discrete dynamical systems on networks [101, 198, 243, 249]. Such ideas have been applied to investigate numerous phenomena, including synchronization of chaotic systems (such as Rössler circuits [96]) on networks. In addition to stability, there are also many studies on the control of dynamical systems on networks [60, 70, 192, 193, 225, 269, 276].

Synchronization, which we discussed using the Kuramoto model as an example in Sec. 3.6, refers to a diverse set of phenomena and is just one form of collective behavior. Additionally, there exist numerous forms of synchronization, which can be rather elaborate. For example, in "cluster synchronization" [246], different sets of nodes synchronize separately, and it is possible to desynchronize the nodes in some of these sets without disturbing other sets of nodes. Additionally, different densely connected communities in networks can synchronize on different time scales [9], and in "chimera states," which have now been observed experimentally in a large variety of systems, some sets of nodes synchronize with each other while other sets consist of nodes that oscillate incoherently [239].

Synchronization and other collective phenomena arise ubiquitously in both natural and engineered systems. For instance, oscillator synchrony on networks can play a role in animal behavior, such as in cattle synchrony [297]. Indeed, it is supposedly beneficial for cattle welfare when cows lie down at the same time. The model for cow behavior in Ref. [297] exhibits an interesting feature that has also been noted more generally for dynamical systems on networks [229]: it is possible that increasing the coupling strength of edges or increasing the number of coupling edges can *lower*, rather than raise, the amount of synchrony. The scenario of reducing synchrony by adding new edges is a direct analog of the Braess paradox [36, 185, 332], a general (and often counterintuitive) phenomenon that has also been studied in traffic systems, power systems, and more. Other forms of collective behavior in animals (e.g., flocking in seagulls and other animals) and many other forms of collective, coordinated, and such dynamics [157, 320] are also studied heavily, although they are often phrased using the language of agent-based models rather than as dynamical processes on networks. A control-theoretic perspective can be useful for studying such dynamics [158, 223, 236, 300, 301]. A particularly well-known and well-studied model for flocking was first introduced about 20 years ago by Vicsek et al. [321], and also recall the voter models from Sec. 3.4. See [248] for a survey of group dynamics on networks using a perspective that combines physics and evolutionary game theory.

Given the natural network description of neuronal systems, it is crucial to investigate simple models of neural signal propagation on networks [11, 91] to illuminate the effects of network structure on dynamics (and, ideally, on functional behavior). For example, it is desirable to understand the synchronization properties of such models in networks with different topologies, and some investigations of Kuramoto models on networks have had neuronal applications in mind. It is necessary, however, to also consider dynamical processes that more directly model neuronal systems. Neural signal models that have been studied on networks include integrate-and-fire models [41], Hindmarsh–Rose oscillators [73], Fitzhugh–Nagumo models [45], and more. In neuroscience applications, it is also important to investigate the effects of delay (e.g., due to propagation along longer neurons) [45, 73].

It is also important to study classical examples of stochastic processes on networks, such as the numerous flavors of random walks [5]. In addition to the intrinsically interesting properties of random walks, they have a wealth of applications—we alluded to one very briefly in our discussion of voter models in Sec. 3.4—that range from clustering "communities" of densely connected nodes in networks [160, 181, 259] to ranking Web pages, sports teams, mathematics programs, and much more [39, 114, 304]. Naturally, many Markov processes other than random walks have also been studied on networks.

As we have discussed at length, a lot of work has examined stochastic processes or ODEs on networks, but comparatively little work has considered PDEs on networks. Thankfully, the amount of scholarship on both linear and nonlinear PDEs on networks is starting to increase (see, e.g., [12, 59, 115, 116, 149, 221]), and there are many exciting avenues to pursue. For example, it is intuitively sensible to study models of vehicular (and other) traffic flow on networks [103], and investigating shock-forming PDEs (like Burger's equation), which have often been employed for models of traffic flow [331], on networks should be really interesting. It is also worthwhile to examine mathematical objects such as delay differential equations, integral equations, and integro-differential equations on networks. They are relevant for many applications, such as for incorporating delay in the propagation of neural signals [11].

Percolation processes and related models have been applied to the study of a multitude of applications [228], such as in the study of biological and social contagions (see Secs. 3.2 and 3.3). In percolation processes, much of the focus is on network connectivity, which also often takes center stage in the study of spreading phenomena. Notably, percolation-based models have also been used for many applications, such as cascading failures in power grids [134] and in finance [102], that also include phenomena (e.g., oscillations) other than spreading. In such applications, which often exhibit nonlocal behavior, one needs to be particularly careful when interpreting results based on percolation models, as they can be misleading. Fascinatingly, a recent investigation of cascading failures in power grids constructed an "influence network" and reported that power outages do propagate locally in that network (even though the propagation is nonlocal in the original network) [134]. For power grids, one type of oscillator model that has been employed resembles the Kuramoto model, except for the very important addition of a term with a second-order time derivative. See, e.g., [332] for a derivation of such a model for a power grid. In financial applications, a family of examples that is related to some of our previous discussions (see Sec. 3.3.1) are models of financial contagions and systemic risk in banking networks in terms of threshold models of complex contagions that include additional complications [44, 102, 130, 145]. One can reduce the simplest such models to threshold models on directed networks, but the incorporation of weighted edges (e.g., to represent the values of interbank loans) is important for the development of increasingly realistic models. It is also essential to incorporate internal dynamics of agents (as pursued in the literature on quantitative finance) and to examine stochastic differential equations on networks of financial entities [40].

Coordination games on networks—a subtopic of the much larger topic of games on networks—are also related directly to threshold models [155]. (See [153, 156] for extensive discussions of an enormous variety of games on networks.) Consider the example of technology adoption, and suppose that there are two choices (i.e., game strategies) A and B that represent, for example, technologies for communicating with friends (e.g., mobile-phone text messaging versus Facebook messaging). The game is played by the nodes on a network, where each node adopts one of the possible strategies. If two neighbors on a network both choose strategy A, then they each receive a payoff of q. If they both choose strategy B, then they each receive $1 - q$. If they choose opposite strategies, then they each receive a payoff of 0 because they cannot communicate with each other. If $q > 1/2$, strategy A is construed as representing the superior technology. Consider a network in which all nodes initially play strategy B, and then a small number of nodes begin to adopt strategy A. If one applies best-response updates to the nodes, they end up adopting strategy A only when enough of their network neighbors have already adopted A (as in the Watts threshold model). Specifically, suppose that m of the k neighbors of a given node are playing strategy A and that the remaining $k - m$ neighbors of the node are playing strategy B. If a node plays A, then its payoff is mq; however, if it plays B, then its payoff is $(k-m)(1-q)$. Comparing the payoffs in the two situations, we see that the node should play strategy B until the fraction m/k of its neighbors who are playing A is at least $1 - q$ [89]. The spreading of technology A thus proceeds as a complex contagion precisely as specified in the Watts threshold model. There exist rigorous mathematical results on the speed of adoption cascades in such coordination games on various network topologies [89, 222]. See Sec. A.3.3 for a discussion of cascades in complex contagions.

Boolean networks are another popular topic in the study of dynamical systems on networks [323]. Each node in a Boolean network, which is directed, is in one of two states (e.g., 0 or 1) at each (discrete) time step, and each node has a function (which is often one that is generated randomly) that defines how the states update depending on the states of the neighboring nodes (i.e., in-neighbors). Boolean networks have been very prominent in the study of genetic-expression dynamics [52, 168, 257,

280], although they have also been used in modeling many other systems. Given a network with N nodes and a Boolean update rule, one can write down a network (sometimes called a "state-transition network") with 2^N nodes and directed edges to represent transition from one N-vector of states to another.

Many other types of dynamical processes have also been studied on networks. For example, there is considerable research on dynamical processes on networks in ecology [19, 240, 267]. Chemical-reaction networks have also been studied for a long time and from many perspectives (see, e.g., [61, 62, 93]).

Chapter 4
General Considerations

Now that we have discussed several families of models as motivation (and because they are interesting in their own right), we present some general considerations for studying dynamical systems on networks. We alluded to several of these ideas in our prior discussions.

4.1 Master Stability Condition and Master Stability Function

In this section, we derive a *master stability condition* (MSC) and a *master stability function* (MSF), which allow one to relate the qualitative behavior of a dynamical system on a network to the structure of the network via eigenvalues of the associated adjacency matrix [243–245]. (One can also, of course, express such results in terms of the eigenvalues of graph-Laplacian matrices.) Computation of matrix spectra (i.e., their sets of eigenvalues) is easy, so relating spectra to dynamics provides a convenient means to obtain necessary and sufficient conditions for the linear stability of equilibria, periodic orbits, or other types of behavior. The use of such techniques is therefore very common in the investigation of phenomena such as synchronization in networks of coupled oscillators [10]. In our discussion, we closely follow (parts of) the presentation in [228], which illustrates these ideas in the context of continuous dynamical systems. As usual, one can consider much more general situations than what we will present [10, 245]. In particular, we present a calculation for equilibrium points (e.g., representing a synchronized state) for a dynamical system, and one can use a variational formulation of an MSF to examine more general types of behavior (including, e.g., synchronization of chaotic dynamics) [245]. Given various assumptions, one can derive a set of variational equations that are the same for all networks, and one can then analyze the generic

© Springer International Publishing Switzerland 2016
M.A. Porter, J.P. Gleeson, *Dynamical Systems on Networks*, Frontiers in Applied
Dynamical Systems: Reviews and Tutorials 4, DOI 10.1007/978-3-319-26641-1_4

variational equations that generate an MSF. In our presentation, we wish only to illustrate a few ideas of why an MSF approach can be very powerful, so we will do a few calculations "by hand."

Let's suppose that each node i is associated with a single variable x_i. We use $\mathbf{x}$ to denote the vector of these variables. Consider the continuous dynamical system

$$\dot{x}_i := \frac{dx_i}{dt} = f_i(x_i) + \sum_{j=1}^{N} A_{ij} g_{ij}(x_i, x_j), \qquad i \in \{1, \ldots, N\}, \tag{4.1}$$

where $\mathbf{A} = [A_{ij}]$ is the adjacency matrix of a network and $g_{ij}(x_i, x_j)$ represents the effect of network neighbors on each others' dynamics. As usual, the equilibrium points for Eq. (4.1) [127, 292] satisfy $\dot{x}_i = 0$ for all nodes i. To determine the local stability of these points, we (of course) do linear stability analysis: let $x_i = x_i^* + \epsilon_i$ (where $|\epsilon_i| \ll 1$) and take a Taylor expansion. We assume that the network represented by the adjacency matrix $\mathbf{A}$ is time-independent, so one can clearly be much more general than what we do for our presentation. For each i, we obtain

$$\dot{x}_i = \dot{\epsilon}_i = f_i(x_i^* + \epsilon_i) + \sum_{j=1}^{N} A_{ij} g_{ij}(x_i^* + \epsilon_i, x_j^* + \epsilon_j)$$

$$= \underbrace{f_i(x_i^*)}_{\eta_1} + \underbrace{\sum_{j=1}^{N} A_{ij} g_{ij}(x_i^*, x_j^*)}_{\eta_2} + \underbrace{\epsilon_i f_i'\big|_{x_i = x_i^*}}_{\eta_3}$$

$$+ \underbrace{\epsilon_i \sum_{j=1}^{N} A_{ij} \frac{\partial g_{ij}}{\partial x_i}\bigg|_{x_i = x_i^*, x_j = x_j^*}}_{\eta_4} + \underbrace{\sum_{j=1}^{N} \epsilon_j A_{ij} \frac{\partial g_{ij}}{\partial x_j}\bigg|_{x_i = x_i^*, x_j = x_j^*}}_{\eta_5} + \ldots, \tag{4.2}$$

where $f_i' := \frac{df_i}{dx_i}$ and we define η_l, for $l \in \{1, \ldots, 5\}$, from the five corresponding terms in Eq. (4.2). Because $\mathbf{x}^*$ is an equilibrium point, it follows that $\eta_1 + \eta_2 = 0$. The terms η_3 and η_4 are linear in ϵ_i, and η_5 is linear in each of the ϵ_j terms. We are doing linear stability analysis, so we neglect all higher-order terms.

To simplify notation, we define

$$a_i := f_i'(x_i)\big|_{x_i = x_i^*},$$

$$b_{ij} := \frac{\partial g_{ij}}{\partial x_i}(x_i, x_j)\bigg|_{x_i = x_i^*, x_j = x_j^*},$$

$$c_{ij} := \frac{\partial g_{ij}}{\partial x_j}(x_i, x_j)\bigg|_{x_i = x_i^*, x_j = x_j^*}. \tag{4.3}$$

We can then write

$$\dot{\epsilon} = M\epsilon + \dots, \tag{4.4}$$

where $M = [M_{ij}]$ and

$$M_{ij} = \delta_{ij} \left[a_i + \sum_k b_{ik}A_{ik} \right] + c_{ij}A_{ij}. \tag{4.5}$$

Assuming that the matrix M has N distinct eigenvectors (which need not always be the case [127], although it will typically be if we are away from a bifurcation point) and is thus diagonalizable, we expand

$$\epsilon = \sum_{r=1}^{N} \alpha_r(t)\mathbf{v}_r, \tag{4.6}$$

where $\mathbf{v}_r$ (with corresponding eigenvalue[1] λ_r) is the rth (right) eigenvector of the matrix M. It follows that

$$\dot{\epsilon} = \sum_{r=1}^{N} \dot{\alpha}_r\mathbf{v}_r = M\epsilon = M\sum_{r=1}^{N} \alpha_r(t)\mathbf{v}_r = \sum_{r=1}^{N} \alpha_r(t)M\mathbf{v}_r = \sum_{r=1}^{N} \lambda_r\alpha_r(t)\mathbf{v}_r. \tag{4.7}$$

Separately equating the linearly independent terms in Eq. (4.7) then yields $\dot{\alpha}_r = \lambda_r\alpha_r$, which in turn implies that $\alpha_r(t) = \alpha_r(0)\exp(\lambda_r t)$. As usual for dynamical systems [127, 292], we obtain local asymptotic stability if $\mathrm{Re}(\lambda_r) < 0$ for all r, instability if any $\mathrm{Re}(\lambda_r) > 0$, and a marginal stability (for which one needs to examine nonlinear terms) if $\mathrm{Re}(\lambda_r) = 0$ for some r and none of the eigenvalues have a positive real part.

As an example, let's consider a (significantly) simplified situation in which every node has the same equilibrium value: that is, $x_i^* = x^*$ for all nodes i. (This arises, for example, in the SI model of a biological contagion.) We also assume that $f_i \equiv f$ for all nodes and $g_{ij} \equiv g$ for all node pairs. These are also major simplifications, but they are employed in the overwhelming majority of studies that use MSFs [10, 245], predominantly because they are convenient. Typically, it is still hard to perform analytical studies of dynamical systems on networks even with these simplifications. After applying the simplifications, we can write

$$f(x^*) + \sum_{j=1}^{N} A_{ij}g(x^*, x^*) = f(x^*) + k_i g(x^*, x^*) = 0, \tag{4.8}$$

---

[1]Note that we previously used the notation λ in Sec. 3.2.1 to represent the transmission rate in the SI model. In this section, we use λ with appropriate subscripts to represent the eigenvalues of A.

where we recall that k_i is the degree of node i. (Recall as well that we are considering
unweighted and undirected networks.) Equation (4.8) implies that either all nodes
have the same degree (i.e., that our graph is "z-regular," where z is the degree)
or that $g(x^*, x^*) = 0$. We do not wish to restrict the network structure severely,
so we suppose that the latter condition holds. It then follows that $f(x^*) = 0$, so
the equilibria of the coupled equations in Eq. (4.1) in this simplified scenario are
necessarily the same as the equilibria of the intrinsic dynamics that are satisfied by
individual (i.e., uncoupled) nodes. This yields a simplified version of the notation
from Eq. (4.3):

$$a_i \equiv a := f'\big|_{x_i=x^*},$$

$$b_{ij} \equiv b := \frac{\partial g}{\partial x_i}\bigg|_{x_i=x_j=x^*},$$

$$c_{ij} \equiv c := \frac{\partial g}{\partial x_j}\bigg|_{x_i=x_j=x^*}. \tag{4.9}$$

We thus obtain

$$\dot{\epsilon}_i = (a + bk_i)\epsilon_i + c \sum_{j=1}^{N} A_{ij}\epsilon_j, \qquad i \in \{1, \ldots, N\}. \tag{4.10}$$

If we assume that $g(x_i, x_j) = g(x_j)$, which is yet another major simplifying
assumption (don't you love how many assumptions we're making?), we obtain

$$\dot{x}_i = f(x_i) + \sum_{j=1}^{N} A_{ij}g(x_j). \tag{4.11}$$

Consequently, $b = 0$ and

$$\dot{\epsilon} = \mathbf{M}\epsilon = (a\mathbf{I} + c\mathbf{A})\epsilon, \tag{4.12}$$

where $\mathbf{I}$ is the $N \times N$ identity matrix.

An equilibrium of (4.12) is (locally) asymptotically stable if and only if all of the
eigenvalues of $\mathbf{M} = a\mathbf{I} + c\mathbf{A} = \mathbf{M}^T$ are negative. (The matrix $\mathbf{M}$ is symmetric, so
all of its eigenvalues are guaranteed to be real.) Let $\mathbf{w}_r$ denote an eigenvector of $\mathbf{A}$
with corresponding eigenvalue λ_r. It follows that

$$(a\mathbf{I} + c\mathbf{A})\mathbf{w}_r = (a + c\lambda r)\mathbf{w}_r \tag{4.13}$$

for all r (where there are at most N eigenvectors and there are guaranteed to be
exactly N of them if we are able to diagonalize $\mathbf{A}$), so $\mathbf{w}_r$ is also an eigenvector of
the matrix $\mathbf{M}$. Its corresponding eigenvalue for the matrix $\mathbf{M}$ is $a + c\lambda_r$. For (local)
asymptotic stability, we thus need $a + c\lambda_r < 0$ to hold for all λ_r. This, in turn,
implies that we need $a < 0$, because the adjacency matrix $\mathbf{A}$ is guaranteed to have
both positive and negative eigenvalues [228]. We thus need (i) $\lambda_r < -a/c$ for $c > 0$

and (ii) $\lambda_r > -a/c$ for $c < 0$. If (i) is satisfied for the most positive eigenvalue λ_1 of $\mathbf{A}$, then it (obviously) must be satisfied for all eigenvalues of $\mathbf{A}$. If (ii) is satisfied for the most negative eigenvalue λ_N of $\mathbf{A}$, then it (obviously) must be satisfied for all eigenvalues of $\mathbf{A}$. It follows that

$$\frac{1}{\lambda_N} < -\frac{c}{a} < \frac{1}{\lambda_1}, \tag{4.14}$$

which becomes much more insightful when we insert the definitions of a and c. We thereby write

$$\frac{1}{\lambda_N} < -\frac{\left.\frac{\partial g}{\partial x_j}\right|_{x_i=x_j=x^*}}{\left.f'\right|_{x=x^*}} < \frac{1}{\lambda_1}. \tag{4.15}$$

The left and right terms in Eq. (4.15), which is called a *master stability condition*, depend only on the structure of a network, and the central term depends only on the nature (i.e., functional forms of the individual dynamics and of the coupling terms) of the dynamics.[2] In our opinion, that's really awesome! Less enthusiastically but even more importantly, it also illustrates that the eigenvalues of adjacency matrices have important ramifications for the qualitative behavior of dynamical systems on networks. Indeed, investigations of the spectra (i.e., set of eigenvalues) of adjacency matrices (and of other matrices, such as different types of graph Laplacians) can yield crucial insights about dynamical systems on networks [228, 285]. These insights have repeatedly been important in the analysis of such systems [10, 242, 313].

Now let's suppose that each node is associated with $\tilde{N}$ variables rather than just one. We now write

$$\dot{\mathbf{x}}_i = \mathbf{f}_i(\mathbf{x}_i) + \sum_{j=1}^{N} A_{ij}\mathbf{g}_{ij}(\mathbf{x}_i, \mathbf{x}_j), \qquad i \in \{1, \ldots, N\}. \tag{4.16}$$

In other words, the variables and functions are now vectors. As before, we do linear stability analysis, and we again derive an equation of the form

$$\dot{\epsilon} = \mathbb{M}\epsilon, \tag{4.17}$$

where ϵ is now a matrix of size $N \times \tilde{N}$ and one can think of $\mathbb{M}$ as a doubly indexed matrix (it's technically a tensor).[3] The component ϵ_{im} denotes the perturbation

---

[2] Although we were able to separate the dependence on structure and dynamics in our example, note that the analysis is more complicated when the equilibrium points are different for different nodes and when considering other types of behavior (e.g., periodic or chaotic dynamics) [244].

[3] See [175] for a discussion of tensors.

(in the linear stability analysis) of the mth variable on the ith node (where $m \in \{1, \ldots, \tilde{N}\}$ and $i \in \{1, \ldots, N\}$). If we assume that the same vector function $\mathbf{f} \equiv \mathbf{f}_i$ describes the intrinsic dynamics on node i and that the coupling function $\mathbf{g} \equiv \mathbf{g}_{ij}$ is the same for all pairs of nodes, then the components $M_{im,jn}$ of $\mathbb{M}$, which we index using a pair of index pairs, are [228]

$$M_{im,jn} = \delta_{ij} a_{mn} + \delta_{ij} k_i b_{mn} + A_{ij} c_{mn}, \qquad (4.18)$$

where

$$a_{mn} := \left. \frac{\partial f_m}{\partial x_n}(\mathbf{x}) \right|_{\mathbf{x}=\mathbf{x}^*},$$

$$b_{mn} := \left. \frac{\partial g_m}{\partial u_n}(\mathbf{u}, \mathbf{v}) \right|_{\mathbf{u},\mathbf{v}=\mathbf{x}^*},$$

$$c_{mn} := \left. \frac{\partial g_m}{\partial v_n}(\mathbf{u}, \mathbf{v}) \right|_{\mathbf{u},\mathbf{v}=\mathbf{x}^*}, \qquad (4.19)$$

and we are using the dummy variables $\mathbf{u} := \mathbf{x}_i$ and $\mathbf{v} := \mathbf{x}_j$ in Eq. (4.19) to prevent confusion. The notation in (4.19) gives the components of a trio of $\tilde{N} \times \tilde{N}$ matrices: $\mathbf{a} = [a_{mn}]$, $\mathbf{b} = [b_{mn}]$, and $\mathbf{c} = [c_{mn}]$.

Let's now assume once again that $\mathbf{g}(\mathbf{x}_i, \mathbf{x}_j) = \mathbf{g}(\mathbf{x}_j)$. This yields

$$\dot{\epsilon}_{im} = \sum_{jn} M_{im,jn} \, \epsilon_{jn} = \sum_{jn} \left(\delta_{ij} a_{mn} + A_{ij} c_{mn} \right) \epsilon_{jn} . \qquad (4.20)$$

As before, we assume for each nodal variable m that we can expand $\epsilon_{im}(t)$ as a linear combination of N distinct eigenvectors $\mathbf{w}_r$ of the adjacency matrix $\mathbf{A}$. We write

$$\epsilon_{im}(t) = \sum_r \alpha_{rm}(t) w_{ri} , \qquad (4.21)$$

and we separately equate the coefficients of the independent eigenvectors $\mathbf{w}_r$ (with corresponding eigenvalues λ_r). The ith component of $\mathbf{w}_r$ is w_{ri}, and the coefficients $\alpha_{rm}(t)$ satisfy the dynamical system

$$\dot{\boldsymbol{\alpha}}_r = (\mathbf{a} + \lambda_r \mathbf{c}) \, \boldsymbol{\alpha}_r(t) , \quad r \in \{1, \ldots, N\} , \qquad (4.22)$$

where λ_r is the rth eigenvalue of the adjacency matrix and the vector $\boldsymbol{\alpha}_r$ has components α_{rm} (where $m \in \{1, \ldots, \tilde{N}\}$).

Define $\sigma(\lambda)$ to be the largest positive real part among the eigenvalues of the $\tilde{N} \times \tilde{N}$ matrix $\mathbf{P} := \mathbf{a} + \lambda \mathbf{c}$. For the dynamics to be (locally) asymptotically stable near the equilibrium point $\mathbf{x}^*$, we require that $\sigma(\lambda_r) < 0$ for all r. The function $\sigma(\lambda)$, which is an example of a *master stability function* [10, 243–245], tends to be easy to evaluate numerically. This is excellent news, because one can then use it readily to obtain interesting insights.

Now that we have slogged through the above calculations, let's review and illustrate what we can learn from MSFs [10, 244, 245]. The idea of studying an MSF (and an MSC) is to have a general way to relate network structure to the behavior of dynamical systems on networks. As we have seen, one can use an MSF or MSC to derive a relation between the spectrum of a network's adjacency matrix (or some other matrix associated with a network) to the stability of some kind of qualitative behavior. Obviously, it is of interest to see how this kind of approach manifests for specific network architectures and for specific functions (and families of functions) that describe the dynamics of the individual nodes (namely, f_i) and those that describe how nodes interact with each other (namely, g_{ij}).

The use of MSFs and MSCs to investigate dynamical systems on networks is widespread and can be very insightful [10, 96, 230, 244, 245, 261, 298]. As discussed in [10, 244], they have been employed to great effect in the examination of the stability of various types of dynamical behavior in coupled oscillators on networks, although there is far from a complete understanding of such phenomena. An MSF can give explicit conditions for how easily a dynamical system on a network can stay in a stable state (such as one that corresponds to synchrony of oscillators associated with the network's nodes), and it is useful to compare the relative ease of remaining stable in different families of networks. In some cases, it is possible to use an MSF to derive necessary conditions for the linear stability of a state in terms of the spectrum of a network's adjacency matrix (or some other matrix associated with it). One can either compute the spectrum numerically or can take advantage of exact analytical expressions, approximate analytical expressions, and bounds about the spectra of appropriate matrices. Using such an analysis, it has been demonstrated that it tends to be very difficult in a "ring" network, in which each node is adjacent to its $2b$ nearest neighbors, to stay in a synchronized stable state for a large class of oscillators and a large variety of ways to couple them [10]. However, by adding a small number of "shortcut" edges (which connect distant nodes in the ring) to such a network [258], it is possible to significantly speed up the return to a stable synchronous state for many types of coupled oscillators [17]. Although one cannot typically use an MSF to fully separate structural and dynamical properties, as we did with the special examples that we discussed above, it is often possible to use an MSF to make broad statements about the stability of states for a large set of dynamical systems without having to individually explore every type of function f_i and g_{ij}. Therein lies the power of an MSF approach. The primary weakness of an MSF approach is that it does not indicate the route towards a stable state, as it is concerned with linear stability [10].

Finally, we note that although our above discussion (and our example calculations, which closely followed the presentation in [228]) included numerous simplifying assumptions for expository convenience, many of them can be relaxed. More general situations (e.g., complex spectra from directed networks, bifurcation phenomena, chaotic behavior, etc.) necessitate more complicated expressions and analysis, but our presentation nevertheless conveys some of the fundamental ideas of an MSF approach. See [10, 244, 245] for further discussion.

4.2 Other Approaches for Studying Dynamical
Systems on Networks

There are also other ways to relate network structure and the behavior of dynamical systems on networks. In this section, we discuss a few of them very briefly.

As we discussed in Sec. 4.1, using an MSF can yield necessary and sufficient conditions for the linear stability of a state (e.g., a synchronized state) of a dynamical system on a network [10]. There are also other approaches for determining conditions for the stability of a state. For example, one way to obtain a necessary condition for stability is to construct spanning trees of the Coates graph (i.e., the network with the self-edges removed) of the Jacobian matrix near that state [75, 90]. Additionally, [24] gives an alternative (and also analytically tractable) approach to an MSF for studying synchronization dynamics. The analysis in [24], which is based on calculating path lengths through edges rather than on the eigenvalues of a matrix, allows one to examine global stability of synchronous states in coupled oscillators with both time-independent and time-dependent interactions. It is thereby also useful for examining coupled oscillators in networks with time-dependent structure [25]. See Chapter 6 for a brief discussion of dynamical systems on time-dependent networks.

Another way to examine the effects of network architecture on dynamics is through the investigation of *coupled-cell networks* [291]. The structure of a coupled-cell network is a graph that indicates how the cells are coupled and which cells are equivalent, and a "multiarrow formalism" [119] allows pairs of nodes to have multiple types of connections between them. This also provides an approach for studying dynamical systems on multilayer networks [28, 173], and a tensorial formalism and the construction of "quotient networks" [291] can be very helpful for exploring general bifurcation phenomena and robust patterns on coupled-cell networks (see, e.g., [118]). Such patterns include various forms of synchrony, and the formalism of coupled-cell networks also allows one to conduct in-depth investigations of synchrony-breaking bifurcations.

Several other mathematical ideas have also been very useful for obtaining insights into the behavior of dynamical systems on networks. For example, isospectral compression, isospectral expansion, and other isospectral transformations can help characterize dynamical systems on networks [42, 43]. Methodology from algebraic geometry is also being employed increasingly to elucidate the qualitative behavior (e.g., the number and type of equilibria) of dynamical systems on networks [61, 62, 163, 210]. More recently, such methodology has also been used to investigate network behavior quantitatively (e.g., with data) using computational algebra and statistics [126, 131, 199]. Methods from computational topology are also being explored increasingly actively in the study of networks (e.g., to analyze spreading processes [303]), and the notion of "large graph limits" have been used to gain insights (which should be compared with the results of mean-field and similar theories) into dynamical systems on networks [208, 209]. Ideas from Morse theory are also worth pursuing [104].

4.3 Discrete-State Dynamics: Mean-Field Theories, Pair Approximations, and Higher-Order Approximations

Several approaches to approximating global (i.e., network-scale) observables have been developed to try to understand the relationship between network structure and local (i.e., node-level) discrete-state dynamics. Given the (discrete-state and possibly stochastic) dynamics, one seeks to accurately predict emergent characteristics of the dynamics (e.g., the number of nodes that are infected with a disease). If an approximation method is amenable to mathematical analysis, it can also be possible to use it to identify bifurcation points or critical parameters that affect the qualitative dynamics. In biological contagion models, for example, it is desirable to estimate the *epidemic threshold* (e.g., via the ratio of transmission rate to recovery rate) that, if equaled or exceeded, enables a disease to spread globally through a networked population.

Analytical approximation approaches vary in their complexity, and there is usually an associated tradeoff in accuracy (as measured, for example, by comparing the prediction from theory with a large-scale Monte Carlo simulation of the dynamics). Theories of mean-field (MF) type are most common, as they can provide reasonable—and in some cases, very high—levels of accuracy and are relatively straightforward to formulate.

We now introduce some typical MF approximation schemes, which we illustrate with an example biological contagion (namely, SI disease-spread dynamics) and a threshold model of a social contagion. We also discuss generalizing beyond mean-field theories.

4.3.1 Node-Based Approximation for the SI Model

We begin by considering node-based approximation schemes for the SI model for biological epidemics (see Sec. 3.2.1). We closely follow the presentation in [228]. The simple dynamics of this example allows one to clearly identify the important approximations.

A *node-based approximation* is one in which a variable x_i is defined for every node i in a network. In a given stochastic simulation of a system, x_i takes the value 1 when node i is infected and the value 0 when it is not. If one then considers an ensemble of stochastic simulations, the dynamics evolves differently in each realization, but one can compute the (time-dependent) expectation $\langle x_i(t) \rangle$ of x_i over all simulations in the ensemble. (In practice, one also needs to compare such an expectation to the sample mean over the simulations that one performs.) To write an equation for the temporal evolution of $\langle x_i \rangle$, we first note that if x_i is 0, then it can change to 1 only when the disease is transmitted (at a rate λ) from an already-infected neighbor of node i. The ensemble-averaged quantities $\langle x_i \rangle$ thus obey the following set of differential equations:

$$\frac{d\langle x_i \rangle}{dt} = \lambda \sum_j A_{ij} \langle (1 - x_i)x_j \rangle , \quad i \in \{1, \ldots, N\} , \tag{4.23}$$

where A_{ij} is the adjacency matrix (so the sum over j has nonzero contributions only from neighbors of node i) and the quantity $\langle (1 - x_i)x_j \rangle$ is the probability (over the ensemble of realizations) that node i is susceptible and node j is infected.

The set of equations in Eq. (4.23) is large—there is one equation for each node in a network—but a more serious issue is that it is not a closed system. To close Eqs. (4.23), we must either approximate the quantity $\langle (1 - x_i)x_j \rangle$ in terms of the variables $\langle x_i \rangle$ or we need to derive an equation for its temporal evolution. For example, by assuming independence—i.e., no "dynamical correlations" [112] between the states of nodes i and j, as we discuss in Sec. 4.3.4—we can write

$$\langle (1 - x_i)x_j \rangle = \langle 1 - x_i \rangle \langle x_j \rangle = (1 - \langle x_i \rangle) \langle x_j \rangle , \tag{4.24}$$

which allow us to solve for $\langle x_i \rangle$. We have just performed a *moment closure* in which we have closed at the first moment to produce a *mean-field theory*. See [177] for a survey about moment closure in numerous situations (including networks).

Alternatively, we can derive

$$\frac{d\langle s_i x_j \rangle}{dt} = -\lambda \langle s_i x_j \rangle + \lambda \sum_{k \neq i} A_{jk} \langle s_i s_j x_k \rangle - \lambda \sum_{l \neq j} A_{il} \langle x_l s_i x_j \rangle , \tag{4.25}$$

where we have written $s_i = 1 - x_i$ for convenience. However, we now have to either approximate the triplet terms (to obtain a so-called *pair approximation* [228], once we also express the pair terms $\langle s_i s_j \rangle$ and $\langle x_i x_j \rangle$ in terms of $\langle s_i x_j \rangle$) or derive dynamical equations for the triplet terms. Of course, if we choose to do the latter, the resulting equations will include quadruplet terms.

As an explicit example of closing Eqs. (4.25) to obtain a pair approximation [228], we use Bayes' theorem to derive the approximations

$$\langle s_i s_j x_k \rangle \approx \frac{\langle s_i s_j \rangle \langle s_j x_k \rangle}{\langle s_j \rangle} \quad \text{and} \quad \langle x_l s_i x_j \rangle \approx \frac{\langle x_l s_i \rangle \langle s_i x_j \rangle}{\langle s_i \rangle} .$$

We also use

$$\langle s_i s_j \rangle = \langle s_i (1 - x_j) \rangle = \langle s_i \rangle - \langle s_i x_j \rangle ,$$

and we thereby derive the closed system of equations

$$
\begin{aligned}
\frac{d\langle s_i x_j \rangle}{dt} &= -\lambda \langle s_i x_j \rangle + \lambda \sum_{k \neq i} A_{jk} \frac{(\langle s_i \rangle - \langle s_i x_j \rangle)\langle s_j x_k \rangle}{\langle s_j \rangle} - \lambda \sum_{l \neq j} A_{il} \frac{\langle x_l s_i \rangle \langle s_i x_j \rangle}{\langle s_i \rangle} \\
&= -\lambda \langle s_i x_j \rangle + \lambda \frac{\langle s_i \rangle - \langle s_i x_j \rangle}{\langle s_j \rangle} \sum_{k \neq i} A_{jk} \langle s_j x_k \rangle - \frac{\lambda}{\langle s_i \rangle} \sum_{l \neq j} A_{il} \langle x_l s_i \rangle \langle s_i x_j \rangle ,
\end{aligned}
$$

$$ (4.26) $$

which, along with Eq. (4.23) and the expression $x_i = 1 - s_i$, constitutes a pair approximation.

The above discussion has given some examples of the moment-closure problems that often arise for stochastic dynamics on networks. See Refs. [177, 218] for much more detail on moment closures, and see Ref. [144] for a discussion of the use of algebraic methods for moment closure.

4.3.2 Degree-Based MF Approximation for the SI Model

Dealing with the issue of moment closure requires truncating a hierarchy of differential equations at some stage. Given the complexity of the equations that arise, it is common to reduce the number of equations by assuming that all nodes of degree k behave in a manner that is dynamically similar. In applying such a scheme, which is sometimes called a *degree-based approximation* [228] or a *heterogeneous mean-field approximation* [318], one is making the assumption that it is reasonable to consider the dynamics (at least as concerns the observables of interest) on a configuration-model network.

Suppose that node i has degree k. We replace $\langle x_i \rangle$ in Eq. (4.23) with a new variable $\rho_k(t)$, which is defined as the fraction of nodes of degree k that are infected at time t. Another way to think of $\rho_k(t)$ is that if, at time t, we choose one node (e.g., node i) from the set of all nodes that have degree k, then the probability that the node is infected is $\rho_k(t)$. We are making the following approximation: we are replacing a quantity that is specific to node i by a quantity that is defined for the entire class of degree-k nodes (and which is assumed to be the same for all nodes in that class). With the MF approximation (4.24), the right-hand side of Eq. (4.23) becomes

$$
\lambda (1 - \langle x_i \rangle) \sum_j A_{ij} \langle x_j \rangle ,
$$

$$ (4.27) $$

and the degree-based approximation allows us to replace $1 - \langle x_i \rangle$ by $1 - \rho_k$. In the same spirit of approximation, we replace the sum over neighbors $(\sum_j A_{ij} \langle x_j \rangle)$ by $k \omega_k$, where $\omega_k(t)$ represents the mean-field approximation for the probability that a given neighbor of i (or, indeed, of any node with degree k, because we do not distinguish between these nodes in the degree-based approximation) is infected.

Because we are assuming that our network is a configuration-model network, the probability that a neighbor of node i (or, indeed, of any specific node) has degree k' is given by

$$\frac{k'}{z}P_{k'},\tag{4.28}$$

where we recall that z is the mean degree. To understand this, consider that when a node (e.g., node i) is part of a network that is connected according to the configuration-model rules described in Chapter 2, it has a larger probability of being adjacent to a high-degree node than to a low-degree node, because a high-degree node has more stubs (i.e., ends of edges) available for selection. Thus, because we know that the neighbors of node i are, by definition, adjacent to node i, it follows that these neighbors are more likely to have a high degree k' than would be expected if we choose a node uniformly at random. (For example, we know that a neighbor of node i cannot have degree 0.) Accounting for this bias gives the $k'P_{k'}$ term in Eq. (4.28), and the denominator z ensures that the probability distribution is correctly normalized, so it sums to 1 when all possible k' values are considered. Returning to the probability ω_k that a neighbor of a degree-k node is infected, we see that the probability that the neighbor has degree k' is given by Eq. (4.28), and the probability that a degree-k' node is infected is $\rho_{k'}(t)$. We can then calculate $\omega_k(t)$ (within the MF approximation) by multiplying these probabilities and summing over all possible values of k'. We thereby obtain

$$\omega_k(t) \equiv \omega(t) = \sum_{k'}\frac{k'}{z}P_{k'}\rho_{k'}(t),\tag{4.29}$$

where we note that ω_k in fact turns out to be independent of k. This is a consequence of the random-linking property of the configuration model, and more generally (e.g., in networks with degree–degree correlations) the values of ω_k need not be the same for every k.

Applying all of our approximation steps to Eq. (4.23) yields an MF degree-based approximation for the SI model:

$$\frac{d\rho_k}{dt} = \lambda k(1-\rho_k)\omega,\tag{4.30}$$

where ω is given by Eq. (4.29). Note that the system (4.30) contains one equation for each degree class k in the network, and typically this number is much smaller than the number N of nodes, so the dimension of the system (4.30) is considerably lower than that of (4.23). To take an extreme example, consider a z-regular network, in which every node has exactly z neighbors. In this case, Eq. (4.30) reduces to a single equation,

$$\frac{d\rho}{dt} = \lambda z(1 - \rho)\rho, \tag{4.31}$$

for the fraction $\rho(t)$ of infected nodes. This is the well-known (and analytically solvable) logistic differential equation that appears in SI models of homogeneous, well-mixed populations [37].

4.3.3 Degree-Based MF Approximation for a Threshold Model

We now derive an MF approximation for the Watts threshold model of a social contagion. (See Sec. 3.3.1 for a discussion of social contagions.) In this section, we give an *ad hoc* derivation that highlights some of the important assumptions of MF approximations. In Appendix A, we discuss a more systematic approach for deriving MF (and other, higher-accuracy) approximations.

Let's consider the Watts threshold model with asynchronous updating. For simplicity, as in Eq. (4.31), we assume that the network is z-regular, but one can readily generalize the derivation to networks with arbitrary degree distributions and degree–degree correlations [29, 241]. Define $\rho(t)$ to be the fraction of nodes that are active at time t. We assume that a given seed fraction $\rho(0)$ of nodes are initially activated, where we choose the seed nodes uniformly at random. To derive an MF approximation, we examine how $\rho(t)$ changes in time. Consider an updating event, in which we have selected a node uniformly at random for a possible change of state. The probability that the chosen node is inactive at time t is $1 - \rho(t)$. We want to calculate the probability that m of the neighbors of the node are active, and we then compare the fraction m/z with the threshold R of the node. The probability that $m/z \geq R$ is the probability that the updating node becomes active, and the activation of a node increases ρ. To continue, we will need to make two assumptions.

In the first of our assumptions, we suppose that all of the neighbors of the selected node are (independently) active with probability $\rho(t)$. This independence assumption is an important one: as we will discuss in assumption (1) of Sec. 4.3.4, such independence cannot be exactly true on networks that contain triangles or other short cycles. However, with this independence assumption, we can write the probability that the chosen node has m active neighbors as the binomial distribution

$$B_{z,m}(\rho) = \binom{z}{m} \rho^m (1 - \rho)^{z-m}, \tag{4.32}$$

because we are considering z neighbors, who are each (independently) active with a probability of $\rho(t)$.

A second important MF assumption arises when we suppose that the probability that the updating node is inactive *and* that it has m active neighbors is given by $[1 - \rho(t)] B_{z,m}(\rho(t))$. To obtain this expression, we multiply the probability $1 - \rho$ of a node being inactive by the probability $B_{z,m}(\rho)$ that it has m active neighbors.

Calculating a joint probability in this way is justified if the two events—in this case, that the "updating node is inactive" and that the "updating node has m active neighbors"—are independent. Consequently, the second important MF assumption (see assumption (2) in our discussion in Sec. 4.3.4) is that the state (active or inactive) of an updating node is independent of the states of its neighbors. This assumption also cannot be exactly correct. For instance, a node whose neighbors are all active is more likely to be active than a node with no active neighbors. The dynamics of the system induce correlations between the states of nodes and their neighbors, but our MF assumptions are ignoring such "dynamical correlations."

Returning to our derivation, we have thus far agreed—we didn't really give you a choice, to be honest, except possibly to look at the more complicated approach in Appendix A—to approximate the probability that the updating node is inactive and has m active neighbors by $[1 - \rho(t)] B_{z,m} (\rho(t))$. We now ask the following question: what is the probability that such a node will become active and thereby increase the active fraction $\rho(t)$? According to the update rule in the Watts model—which becomes equivalent to the Centola–Macy update rule on a z-regular network—the node will become active if the fraction m/z of active neighbors equals or exceeds its threshold R. Recall that we chose R from a predefined distribution, so we need to determine the probability that a threshold value R drawn from this distribution is less than or equal to the value m/z. This probability is given by

$$\text{Prob}\left(R \le \frac{m}{z}\right) = C\left(\frac{m}{z}\right), \tag{4.33}$$

where C is the cumulative distribution function (CDF) of the thresholds.

Putting together the probabilities that we calculated above and summing over the possible values of m, we obtain the following MF approximation for the fraction $\rho(t)$ of active nodes:

$$\frac{d\rho}{dt} = (1 - \rho) \sum_{m=0}^{z} B_{z,m}(\rho) C\left(\frac{m}{z}\right). \tag{4.34}$$

Equation (4.34) is a nonlinear ODE for $\rho(t)$, with an initial condition of $\rho(0)$, and one can easily solve it numerically to obtain the time-dependent MF prediction for $\rho(t)$. In Fig. 4.1, we show a few examples, in which we compare the solutions of the MF equation (green dash-dotted curves) with ensemble averages of direct (stochastic) numerical simulations of the Watts model (black symbols). See Sec. 5.1 for a description of how to perform such numerical simulations. Although the MF predictions in Fig. 4.1 are qualitatively correct for the better-connected networks (with $z = 6$ and $z = 7$), the quantitative agreement tends to be poor. Moreover, on low-degree networks (e.g., $z = 4$), the MF approximation is even qualitatively incorrect. It predicts that the social contagion spreads through the whole network,

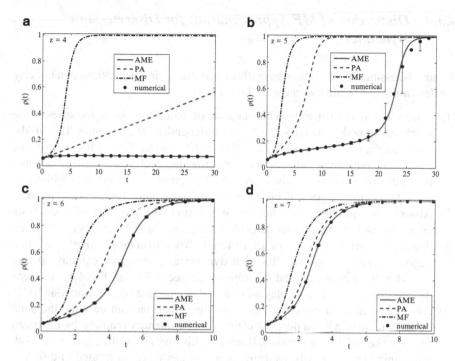

Fig. 4.1 Fraction $\rho(t)$ of infected nodes in a Watts threshold model on z-regular random graphs (for which the degree distribution is $P_k = \delta_{k,z}$) for $z \in \{4, 5, 6, 7\}$. We show the results of the mean-field (MF) approximation (4.34) using the green dash-dotted curves, and we also show the results of pair approximation (PA) and approximate master equation (AME) schemes. (We discuss these latter two approximations in Appendix A.) The initial fraction of infected nodes is $\rho(0) = 0.055$ in all cases, and all nodes in each network have the same threshold level of $R = 2/z$ (so that a node activates if 2 or more of its neighbors are active). We show the results of numerical simulations (see Chapter 5) for networks with $N = 10^5$ nodes, and we use a time step of $dt = 10^{-5}$. The black symbols show the means over 24 realizations, and the error bars indicate one standard deviation above and below the mean. Each realization uses the same network from the random graph ensemble, but different nodes are infected initially (and the update order is also different).[4] Because we are examining large z-regular random graphs, the dynamics are essentially the same for any network from a given random-graph ensemble.

whereas in fact only a very small fraction of nodes ever adopt the contagion. To address this situation, one can use higher-accuracy approximations (as we describe in Appendix A), albeit at the cost of increased complexity of the dynamical system and of the derivation.

---

[4] When calculating a sample mean using numerical simulations of a dynamical system on a network (or a family of networks), there are several possible sources of stochasticity: (1) choice of initial condition, (2) choice of which nodes to update (when considering asynchronous updating), (3) the update rule itself, (4) parameter values that are used in an update rule, and (5) selection of particular networks from a random-graph ensemble. Some or all of these sources of randomness can be present when studying dynamical systems on networks, and (when possible) it is also desirable to compare the sample means to ensemble averages (i.e., expectations over a suitable probability distribution).

4.3.4   Discussion of MF Approximation for Discrete-State Dynamics

As our discussions in this section have illustrated, the main assumptions that underly the derivation of MF theories are as follows [112]:

(1) Absence of local clustering: When the state of node i is updated, the states of the neighbors of node i are considered to be independent of each other. This holds, for example, when one makes a *locally tree-like* assumption on the structure of a network. With such a structural assumption, the cycles in a network become negligible as $N \to \infty$, so there are very few pairs of connected neighbors to consider in the first place. (This motivates the terse name for the assumption.)

(2) Absence of *dynamical correlations*: When updating the state of node i, its own state and the state of its neighbors are assumed to be independent. Such dynamical correlations are rather different from *structural correlations* like degree–degree correlations. Note that dynamical correlations can have strong effects even in z-regular random graphs (see Sec. 4.3.3 and Fig. 4.1), whereas degree–degree correlations play no role in this situation by construction [107].

(3) Absence of modularity: It is usually assumed that one can describe the state of every node of a given degree k using a single quantity (namely, by the mean over the class of degree-k nodes). However, this may not be the case if a network has significant modularity, as degree-k nodes can often be located in different communities, and there can also be other sources of diversity among the degree-k nodes.[5]

One can capture dynamical correlations in part by using generalizations of MF theories that incorporate information on the joint distribution of node states at the ends of a random edge in a network. Such theories are often called *pair approximations* (PA) and are more complicated to derive than MF theories [218]. However, they also tend to be more accurate than MF theories. One can also include the dynamics of triplets, quadruplets, and as many nodes as one wants by considering more general *motif expansions* [71, 218, 302].

It is also possible to derive very accurate approximations using compartmental models in which one considers the states of all neighbors when updating a node. Such approaches are expensive computationally, but it has been demonstrated that they can yield high accuracy for models of biological contagions [191, 201]. They have also been successfully generalized for a wide class of binary-state models [108]. Reference [108] also illustrates how to systematically derive PA and MF theories from higher-order compartmental models. See Appendix A for some technical details.

---

[5]It may also be useful to develop analogous class-based approximations that use structural characteristics other than degree.

Despite the fact that MF assumptions (1)–(3) are violated for many (and perhaps most?) real-world networks, MF theories often give reasonably accurate predictions of global dynamics—at least for well-connected networks and when the dynamical system under consideration is far from any bifurcation points. However, for sparse networks and for accurate bifurcation analysis, one often needs to work harder to achieve high levels of accuracy. This motivates a move beyond pair approximations to methods (such as motif expansions) that include more information on the states of the neighbors of each node [31, 71].

MF theories and their generalizations provide analytical approaches for reducing the dimensionality of a dynamical system on a network via an approximation scheme. Alternatively, one can take a philosophically similar approach (for both dynamics on networks and dynamics of networks) using only computations. See Refs. [32, 262] for examples of such an "equation-free" approach.

4.4 Additional Considerations

As we discuss in Appendix A, there exist rather sophisticated approximation methods for binary-state dynamics on configuration-model networks. Moreover, for monotonic binary dynamics, there has also been significant progress in extending theories to networks with degree–degree correlations, clustering, modularity, and multilayer structures. Indeed, degree–degree correlations have been incorporated into many MF and PA methods.

However, despite this progress, much remains to be done. Moving beyond binary-state dynamics is a significant challenge. There has been notable progress in some dynamical systems that include three states—particularly when the dynamics remains monotonic (i.e., transitions between states still occur only in one direction). Examples include the multi-stage complex contagion model of social influence in Ref. [212] and high-accuracy approximations for SIR disease spread [191, 215]. It would be very interesting—though also rather challenging—to extend the compartmental AME approach to dynamics with more than two states and to subsequently use the results of such analysis to derive PA and MF approximations like (A.4) and (A.6) in a systematic fashion. Other open problems include developing high-accuracy approximations of nodal dynamics with a continuum of states—e.g., with differential equations at each node, as in the Kuramoto model of coupled oscillators [147]—and developing a better understanding of the novel features that were observed in [76] for non-monotonic binary dynamics with synchronous updates. Given observed binary-state dynamics, one can also examine the inverse problem of trying to reconstruct network architecture [189].

Chapter 5
Software Implementation

In this chapter, we briefly discuss practical issues related to simulating dynamical systems on networks.

5.1 Stochastic Simulations (i.e., Monte Carlo Simulations)

As we noted in Sec. 3.5, it is relatively straightforward to implement a Monte Carlo simulation of stochastic discrete-state dynamics using equal-length time steps. (See our discussion in Sec. A.1.1 as well.) The special case of monotonic threshold dynamics is especially simple, so we will now describe it briefly in the context of the Watts threshold model. See [138] for discussions of fast algorithms for epidemic models on networks, and see [30, 319] for Gillespie-type algorithms in which the time steps do not have equal lengths.

Given an adjacency matrix $\mathbf{A}$ and an $N \times 1$ vector $\mathbf{v}$ that stores the states of each node (at a given time, $v_i = 0$ if node i is susceptible and $v_i = 1$ if node i is infected), we calculate the number m_i of infected neighbors of each node i using matrix multiplication:

$$\mathbf{m} = \mathbf{A}\mathbf{v}. \tag{5.1}$$

Similarly, the degree k_i of each node i is the ith element of the vector $\mathbf{k}$ defined by

$$\mathbf{k} = \mathbf{A}\mathbf{1}, \tag{5.2}$$

where $\mathbf{1}$ represents the $N \times 1$ vector $(1, 1, 1, \ldots, 1)^T$. At each time step in an asynchronous updating scheme (see Sec. 3.5), a node i is chosen uniformly at random for updating, and v_i is set to 1 if m_i/k_i equals or exceeds the threshold R_i of node i.

© Springer International Publishing Switzerland 2016   47
M.A. Porter, J.P. Gleeson, *Dynamical Systems on Networks*, Frontiers in Applied
Dynamical Systems: Reviews and Tutorials 4, DOI 10.1007/978-3-319-26641-1_5

One then uses the updated state vector **v** to calculate the updated vector **m** of infected neighbors from Eq. (5.1), and the simulation continues to the next time step. One can terminate the temporal evolution when the condition

$$\frac{m_i}{k_i} < R_i \quad \text{for all nodes } i \text{ with } v_i = 0 \tag{5.3}$$

is satisfied, as this implies that no further susceptible nodes can be infected and thus that the system has reached a steady state. One can show [109] that the steady state of monotonic threshold dynamics is independent of whether one employs asynchronous or synchronous updating, so one can accelerate the algorithm that we just described by updating all nodes simultaneously in each time step if one is interested only in the steady state. (Naturally, it is often the case that one wishes to explore phenomena other than a steady state [212].)

5.2   Differential-Equation Solvers for Theories

OCTAVE/MATLAB code for implementing and solving the systems of ODEs that arise from various approximation schemes for stochastic binary-state dynamics is available at http://www.ul.ie/gleesonj/solve_AME. In this code, a user specifies as inputs the degree distribution P_k of a network, the initial fraction $\rho(0)$ of infected nodes, and the transition rates of the dynamics. (See Sec. A.1.1 for a description of these rates.) The code then automatically implements the approximate master equations (AME) (see Ref. [108] and the discussion in Appendix A), the pair-approximation equations in Eq. (A.6), and the mean-field equations in Eq. (A.4) as systems of ODEs and solves them using standard numerical techniques for differential equations. Results are output as plots that show the infected fraction $\rho(t)$ of nodes and the (time-dependent) fraction of edges that connect susceptible nodes to infected nodes.

Additionally, Gerd Zschaler and Thilo Gross have posted software [335] for simulating adaptive networks at http://www.biond.org/node/352. In an adaptive network, recall that a dynamical system on the network is coupled to structural changes of the network [124, 272]. See Chapter 6 for a general discussion of dynamical systems on time-dependent networks.

Chapter 6
Dynamical Systems on Dynamical Networks

The study of dynamical systems on time-dependent (i.e., "temporal" or "dynamical") networks has become extremely popular recently, but there are also much older quantitative studies of such situations. For example, Farmer et al. [92] and Bagley et al. [13] used such a framework more than two decades ago in studies of chemical reactions. Moreover, even in the early part of the 20th century, biostatistician Ronald Fisher posited that one could describe the seemingly random fluttering of a colony of butterflies as a dynamical network of information [97].

In prior chapters, we focused on dynamics that occur on time-independent networks or on ensembles of such networks. The presence of temporal dynamics of network nodes and/or edges evokes crucial modeling issues. One needs to ponder whether one should study a dynamical system on a temporal network, study only a temporal network, study a dynamical system on a time-independent network (or on an ensemble of such networks), or perform multiple such studies [139]. There are also numerous ways to "project" from a temporal network to a time-independent network (or an ensemble of such networks), as such a transition involves issues about the interactions (e.g., time scales, durations, and continuous versus discrete).

If one is considering a single time-independent (i.e., "static") network, then one is assuming that the network's structure does not change on the time scale of the nodal dynamics or at least that it changes so little that it is permissible to pretend that it is time-independent.[1] If one is studying a dynamical process on an ensemble of time-independent random graphs, there are two main possibilities:

- We are in the same situation as above as concerns the balance of time scales (so we have an ensemble of static networks), and the randomness is employed because of uncertainties about the network structure. (For example, perhaps one

---

[1]In the physics literature, such situations with extremely slow structural dynamics are sometimes called "quenched," because the networks are almost frozen.

© Springer International Publishing Switzerland 2016

M.A. Porter, J.P. Gleeson, *Dynamical Systems on Networks*, Frontiers in Applied Dynamical Systems: Reviews and Tutorials 4, DOI 10.1007/978-3-319-26641-1_6

believes in a certain situation that only degree distribution is important, so one
fixes the degree distribution and randomizes everything else.)
- The network structure changes on such a fast time scale that it is reasonable
  only to use a random-graph ensemble (rather than something more specific)
  to describe its properties. That is, instead of an ensemble of "static" networks,
  one uses a description that takes the form of a statistically stationary probability
  distribution,[2] as only appropriate averaged properties of the network are reliable.

There are many situations in which a network itself can be dynamic, with edges
(or edge weights) and/or nodes changing in time, perhaps in response to a dynamical
system associated with the nodes. In general, three basic situations are possible
[139, 242, 252]:

- The dynamics of a network are much faster than the dynamical system that one
  is examining on that network. In this case, it is reasonable (at least to a first
  approximation) to assume that the states of the network nodes are fixed (e.g.,
  a "susceptible" node will not change to being "infected") and to consider only
  the dynamics of the underlying network. (One can also make a similar comment
  about the states of edges if that is what one is studying.) However, if the dynamics
  of the network structure are too fast, then it may be necessary (as we indicated
  above) to consider ensembles of time-independent networks, because only the
  measurement of suitable averaged properties is appropriate.
- A dynamical system on a network evolves on a much faster time scale than
  the dynamics of the network itself. In this case, it is reasonable (at least to a
  first approximation) to assume that the network is time-independent. However, if
  the states of the nodes—or edges, if one is studying their states—are changing
  too rapidly, then it may only be reasonable to consider statistical ensembles
  of the states, because only the measurement of suitable averaged properties is
  appropriate.
- The dynamics on a network and the dynamics of the network itself operate
  on comparable time scales, so it is not reasonable to ignore either of them.
  Such networks are sometimes called "adaptive networks" [124, 272]. This
  terminology emphasizes the coupling between the dynamics on the networks and
  the dynamics of the networks.

The dynamics of networks can have a profound impact on dynamical systems
on networks, and this has now been explored in numerous papers. See, e.g.,
[25, 26, 135, 136, 139, 141–143, 150, 164, 166, 203, 219, 223, 235, 251–253, 274,
281, 288, 299, 301, 309, 322] and a multitude of other references. The bursty—
and, more generally, non-Poisson—nature of interactions in temporal networks
affects both (1) the effective weights of time-independent interactions derived
from aggregating over time-dependent interactions and (2) the behavior of dynamical processes on time-dependent networks versus processes on their associated

---

[2]In the physics literature, such an idea is invoked to justify certain approximations in models, and
the word "annealed" is sometimes used to describe such a situation.

time-independent networks [30, 135, 136, 162, 252, 264]. It is also important to develop computational techniques, such as Gillespie algorithms, for dynamical processes on temporal networks [204, 319].

A good example of a dynamical system on a time-dependent network [139, 141, 142] is an adaptive SIS model [125], in which a susceptible node can break any edge it has to an infected node and rewire to a randomly chosen susceptible node. Compartmental approaches have been applied successfully to this model [201], and they agree very well with the results of numerical simulations. Similar approaches were also used to study a two-opinion adaptive voter model in Ref. [87], which built on the model in [140]. (This model was generalized to $n > 2$ opinions in [279] and to include mechanisms that reinforce local clustering in [200].) In [87], an edge of a network is chosen uniformly at random at each time step. If the opinions of the two nodes attached to the edge are different, then one node imitates the opinion of the other with a probability of $1 - \alpha$. However, with probability α, the edge is broken and one node instead makes a new connection to another node (chosen, depending on the variant of the model, either at random from the whole network or at random from those who hold the same opinion as the node). On a finite network, such adaptive voter dynamics eventually results in a steady state in which a network can contain disconnected components, where each component contains only nodes who share the same opinion. Adaptive voter dynamics provide a challenging testbed for approximation methods. The review [71] carefully examined an adaptive voter model and concluded that none of the approximation schemes that it tested were able to give fully satisfactory results in all regions of parameter space.

A nice way to examine dynamical systems on temporal networks is to use so-called *activity-driven networks* [251]. One constructs an "activity potential" for each agent in such a network to encapsulate the number of interactions that it performs in a characteristic time window. One thereby attempts to characterize interactions between agents, and activity rates can come either from specified functions or from empirical data. Using the activity potentials (which can be different for different agents), one can construct an instantaneous temporal description of network dynamics. This approach has led to insights on phenomena such as the emergence of strong ties in temporal communication networks [167] and how to control contagions in temporal networks [194].

It can also be useful and interesting to take a statistical perspective on dynamical systems on temporal networks (see, e.g., [283, 284]). For example, one can use a time-dependent generalization of exponential random graph models (ERGMs) [196] and thereby study a temporal ERGM (TERGM) [72].

When studying dynamical systems on time-dependent networks, there is a lively debate about whether dynamics of networks slow down dynamical processes or speed them up [139, 141, 142, 164, 166, 203, 253, 274, 299] (and, naturally, the answer is different for different dynamical systems), and it is desirable to examine such effects on a wide variety of dynamical systems. For example, bursty communication patterns can either speed up or slow down adoption speed in threshold models, and temporal-network generalizations of the Watts (i.e., fractional threshold) and Centola–Macy (i.e., absolute threshold) complex-contagion models exhibit interesting differences [164, 299].

Chapter 7
Other Resources

We now list several references that can complement the present monograph, and we indicate particular directions for which we think that you will find them helpful. Obviously, many other resources (e.g., several books) are also available.

- Reference [294] is a friendly introduction to dynamical systems on networks. It surveys the state of the field in 2001, so it is out of date in many respects. (Network science is a young, immature field that advances quickly.)
- Reference [226] is a more recent (but very short) expository article about dynamical systems on networks.
- Reference [318] is a survey article that discusses dynamical systems on networks. It is not very technical, but it provides a big-picture overview of several areas.
- Reference [228] is a very good textbook on network science, and some of its later chapters cover dynamical systems on networks. For example, some of our discussion of general considerations in Chapter 4 followed part of the presentation in this book, which also has thorough discussions of processes like percolation and biological contagions. Reference [228] also discusses the use of mean-field theories and related techniques for studying such dynamical processes, and we drew on some of those discussions for portions of our presentation. In particular, in Sec. 4.3.1, we largely followed the presentation in [228].
- Reference [174] is a textbook on networks that takes a statistical perspective. It includes material on dynamical systems on networks.
- Two of the classical review articles in network science are [227] and [27]. However, because network science advances quickly, they are somewhat out of date.
- Reference [18] is a book devoted to dynamical systems on networks from a physics perspective.
- Reference [234] is a recent survey of spreading processes on networks from a control-theoretic perspective.
- The survey article [38] examines flows on networks from a mathematical perspective.

© Springer International Publishing Switzerland 2016   53
M.A. Porter, J.P. Gleeson, *Dynamical Systems on Networks*, Frontiers in Applied Dynamical Systems: Reviews and Tutorials 4, DOI 10.1007/978-3-319-26641-1_7

- Reference [270] is a review article on recent advances in percolation theory.
- Reference [242] is a recent, extensive review of contagions on networks. It focuses especially on biological contagions, but it also includes discussions of other spreading processes.
- Reference [99] is a book about biological contagions on networks, although there is also some discussion of the spread of information.
- Reference [218] is a review of contagions on networks that focuses on moment-closure methods and discusses the challenges of generalizing methods from configuration-model networks to more complicated random-graph ensembles.
- Reference [48] is a review of social dynamics from the perspective of statistical physics.
- Reference [156] is a survey about games on networks.
- Reference [8] reviews synchronization on networks, although there has been a lot of work on this topic since it was published.
- Reference [192] is a recent review of control in networks, and Ref. [225] is a recent survey of the topic.
- Reference [11] is a review of oscillatory dynamics on networks in the context of neuroscience. In addition to its discussions of neuroscientific phenomena and models, it also discusses numerous broader theoretical issues that are important for studying coupled oscillators on networks.
- References [28, 173, 186, 271, 325] all include extensive discussions that review work on various types of dynamical processes on multilayer networks. The body of such work is exploding rapidly, and these five articles offer different perspectives.
- The survey articles [139, 141] and the edited collection [142] discuss a wealth of research on time-dependent networks (so-called "temporal networks"), including many topics in dynamical processes on temporal networks.
- The tutorial article [71] discusses moment-closure approximations for discrete adaptive networks, in which a discrete dynamical system is coupled to a network that changes in discrete time. References [124, 272] review research on adaptive networks.
- Reference [80] is a review article that discusses analytical methods to study critical phenomena in networks. It includes discussions of phase transitions, percolation, synchronization, and many other phenomena.

Chapter 8
Conclusion, Outlook, and Open Problems

In this monograph, we have given a tutorial for studying dynamical systems on networks. By reading our tutorial, you should now have a reasonable understanding of (1) why it is interesting and desirable to study dynamical systems on networks; (2) several of the popular families of problems and models; (3) basic considerations about dynamical systems on networks; (4) the use and range of validity of techniques for analytical approximations such as mean-field theories, pair approximations, and higher-order motif expansions; and (5) time-scale issues and challenges for investigating dynamical systems on time-dependent networks. We have also given pointers to software implementations for direct numerical simulations and for solving the systems of equations that result from the aforementioned approximation methods. As our monograph is a tutorial rather than a literature review, there is a lot that we haven't covered, and we strongly encourage you to scour the literature for interesting problems to study and generalize. The primary purpose of our tutorial is to equip you with the background knowledge to be able to do so successfully. We have also provided numerous references to get you started.

Before saying "goodbye," it is also worth commenting on some of the particularly challenging problems that are available. For example, although one can study problems by purely computational means, we believe that it is desirable (when possible) to try to develop analytical techniques in order to gain insights on these problems. In network science, mathematically rigorous results tend to be rare—see, e.g., [85, 86]—but approximations and heuristic techniques have been employed on many toy problems. A key goal is to find the "next easiest" sets of dynamical systems on networks to study using more general versions of these techniques and to use them to help develop these methods further (and to devise new methods). Additionally, many of the dynamical processes that have been investigated are either some type of percolation or fairly (or even very) closely related to percolation, and it is also important to move beyond these types of models.

© Springer International Publishing Switzerland 2016   55
M.A. Porter, J.P. Gleeson, *Dynamical Systems on Networks*, Frontiers in Applied
Dynamical Systems: Reviews and Tutorials 4, DOI 10.1007/978-3-319-26641-1_8

There are many open issues in the study of dynamical systems on time-dependent networks. One needs to examine the balance of time scales for dynamics on networks versus the dynamics of the networks themselves [139, 252], and the effects that this has on the validity and choices of analytical techniques to use is also very important. Most of the studied situations with both dynamics on networks and dynamics of networks tend to be rather unrealistic, so this is an area that is particularly ripe for further study.

Finally, many networks are *multiplex* (i.e., include multiple types of edges) or have other *multilayer* features [28, 173]. The existence of multiple layers on which different dynamical processes can occur and the possibility of both structural and dynamical correlations between layers offers another rich set of opportunities in the study of dynamical systems on networks. The investigation of dynamical systems on multilayer networks is only in its infancy, and this area is also loaded with a rich set of problems [28, 173, 186, 271].

Goodbye.

Appendix A
Appendix: High-Accuracy Approximation Methods for General Binary-State Dynamics

A.1 High-Accuracy Approximations for Binary-State Dynamics

To illustrate some general concepts for how network structure affects dynamics, let's examine a class of stochastic binary-state dynamics on configuration-model networks. Recall from Chapter 2 that a network in a configuration-model ensemble is specified by a degree distribution P_k but is otherwise maximally random [33, 228]. Pairs of stubs (i.e., ends of edges) are connected to each other uniformly at random, so no degree–degree (or other) correlations are input, and taking the $N \to \infty$ limit guarantees negligible clustering (if we assume, e.g., that the second moment of P_k remains finite as $N \to \infty$). In this section, we discuss a general class of binary-state dynamics and then examine approximation methods at mean-field, pair-approximation, and higher-order levels [218]. We choose to focus on approximations that yield deterministic systems in the form of ODEs. Deterministic approximations are valid when, for example, the number of initially infected nodes (i.e., the seed size of infections) is sufficiently large so that one can neglect stochastic (i.e., realization-to-realization) fluctuations. Otherwise, it is necessary to use other methods, such as ones that are based on branching processes [111, 113, 231–233].

A.1.1 Stochastic Binary-State Dynamics

In a binary-state dynamical system, each node is in one of two states at any time. For convenience, we usually refer to these states as *susceptible* (i.e., of type *S*) and *infected* (i.e., of type *I*) in our discussion. However, such states could actually mean

© Springer International Publishing Switzerland 2016 57
M.A. Porter, J.P. Gleeson, *Dynamical Systems on Networks*, Frontiers in Applied Dynamical Systems: Reviews and Tutorials 4, DOI 10.1007/978-3-319-26641-1

"spin up" versus "spin down" magnetic dipoles in an Ising spin model, "inactive" versus "active" nodes in a social system, and so on. We presented many examples of binary-state dynamics in Chapter 3. See, for example, our discussions of percolation models in Sec. 3.1, biological contagions in Sec. 3.2, and social contagions in Sec. 3.3.

In the present section, we will consider asynchronous updating and local rules for node updates. In other words, the transition rates depend only on the state of a node and on the states of its immediate neighbors in a network. Consequently, an updating node that is susceptible becomes infected with a probability of $F_{k,m} dt$, where k is the node's degree and m is the number of its neighbors that are infected. We refer to the function $F_{k,m}$ as the *infection rate*. Similarly, an updating node that is infected becomes susceptible with a probability of $R_{k,m} dt$, and we refer to $R_{k,m}$ as the *recovery rate*.[1] (Note that one can also consider functions $F_{k,m}$ and $R_{k,m}$ that depend on network diagnostics other than degree, but we restrict ourselves to degree.)

It is straightforward to implement these update rules using Monte Carlo simulations. During a small time step dt, a node i of degree k with m infected neighbors is endowed with probability $\pi_i = F_{k,m} dt$ (if the node is susceptible) or probability $\pi_i = R_{k,m} dt$ (if it is infected) of changing its state. By drawing random numbers r_i from the uniform distribution on $[0, 1]$ and comparing the probabilities recorded for each node, the nodes that change states are the nodes i for which $r_i \leq \pi_i$. The time step dt needs to be sufficiently small so that the number of nodes that change state in a single time step is a small fraction of N. As discussed in Sec. 3.5, it is common to make dt sufficiently small so that a single node is updated.

We now consider some examples of binary-state dynamics of the type that we just described. We begin with the classical SIS mode of disease spread. We endow this model with a transmission rate of λ and a recovery rate of μ. In a small time interval dt, each susceptible node has probability λdt of being infected by each of its infected neighbors. The probability of a susceptible node becoming infected during the time interval dt is then $m dt$ as $dt \to 0$ [see Eq. (3.1)]. We thus identify the infection rate for SIS dynamics as

$$F_{k,m}^{\text{SIS}} = \lambda m, \tag{A.1}$$

which is linear in the number m of infected neighbors. In the SIS model, each infected node can recover to the susceptible state at a constant rate μ. This yields

$$R_{k,m}^{\text{SIS}} = \mu, \tag{A.2}$$

which is independent of m.

As another example, consider the standard voter model (see Sec. 3.4). In this example, the labels "susceptible" and "infected" refer to the two opinions. At

---

[1] The notation $R_{k,m}$ should not be confused with the threshold value R_i of node i in the Watts model and other threshold models.

each time step, one node is chosen uniformly at random for updating (during the interval $dt = 1/N$), and this node copies the state of one of its neighbors (chosen uniformly at random). Therefore, a degree-k node that has m infected neighbors has probability m/k of copying an infected neighbor and probability $(k - m)/k$ of copying a susceptible neighbor. This yields

$$F_{k,m}^{\text{Voter}} = m/k \quad \text{and} \quad R_{k,m}^{\text{Voter}} = (k - m)/k. \tag{A.3}$$

Within this general framework, one can consider a wide class of well-studied binary-state dynamics. Table I in Ref. [108] lists the infection rates $F_{k,m}$ and recovery rates $R_{k,m}$ for models such as the Bass model for the spread of innovations [20, 77], the Ising spin model, the majority-vote model [67, 190], and threshold opinion models. This unified perspective on stochastic binary-state dynamics also allows general derivations of mean-field and pair approximations from a higher-order approximation scheme.

A.2   Approximation Methods for General Binary-State Dynamics

Reference [108] derived equations for degree-based mean-field (MF) and pair-approximation (PA) theories for general stochastic binary-state dynamics, which are defined in terms of an infection rate $F_{k,m}$ and a recovery rate $R_{k,m}$. The MF equations are

$$\frac{d\rho_k}{dt} = -\rho_k \sum_{m=0}^{k} R_{k,m} B_{k,m}(\omega) + (1 - \rho_k) \sum_{m=0}^{k} F_{k,m} B_{k,m}(\omega), \tag{A.4}$$

where $B_{k,m}(\omega)$ denotes the binomial term $\binom{k}{m} \omega^m (1 - \omega)^{k-m}$ and we recall that $\rho_k(t)$ is the fraction of nodes of degree k that are infected at time t. We obtain the quantity $\omega(t)$ in terms of $\rho_k(t)$ using Eq. (4.29). For example, if we consider the SI model for disease spread, then $F_{k,m} = \lambda m$ and $R_{k,m} = 0$. Using the binomial sum

$$\sum_{m=0}^{k} m B_{k,m}(\omega) = k\omega \tag{A.5}$$

in Eq. (A.4) yields the MF theory that we derived in Eq. (4.30).

   If a network has non-empty degree classes from $k = 0$ up to some cutoff k_{max}, then Eq. (A.4) consists of a closed system of at most $k_{\text{max}} + 1$ nonlinear differential equations. (If some of those degree classes are empty, then there will be fewer differential equations.) Standard numerical methods for simulating differential equations enable one to solve these equations efficiently (see Chapter 5).

One can obtain better accuracy than the MF equations in Eq. (A.4) using the PA equations

$$\frac{d\rho_k}{dt} = -\rho_k \sum_{m=0}^{k} R_{k,m} B_{k,m}(q_k) + (1 - \rho_k) \sum_{m=0}^{k} F_{k,m} B_{k,m}(p_k) ,$$

$$\frac{dp_k}{dt} = \sum_{m=0}^{k} \left[p_k - \frac{m}{k} \right] \left[F_{k,m} B_{k,m}(p_k) - \frac{\rho_k}{1 - \rho_k} R_{k,m} B_{k,m}(q_k) \right] + \beta^s (1 - p_k) - \gamma^s p_k ,$$

$$\frac{dq_k}{dt} = \sum_{m=0}^{k} \left[q_k - \frac{m}{k} \right] \left[R_{k,m} B_{k,m}(q_k) - \frac{1 - \rho_k}{\rho_k} F_{k,m} B_{k,m}(p_k) \right] + \beta^i (1 - q_k) - \gamma^i q_k$$

$$(A.6)$$

for the variables $\rho_k(t)$, $p_k(t)$, and $q_k(t)$. The rates β^s, γ^s, β^i, and γ^i are determined from these variables. For example,

$$\beta^s = \frac{\sum_k P_k (1 - \rho_k) \sum_m (k - m) F_{k,m} B_{k,m}(p_k)}{\sum_k P_k (1 - \rho_k) k (1 - p_k)} \tag{A.7}$$

is the rate at which SS edges (i.e., edges that are attached to two nodes of type S) become SI edges. See Ref. [108] for details. The quantity $p_k(t)$ [respectively, $q_k(t)$] is the probability that a randomly chosen (in particular, chosen uniformly at random) neighbor of a susceptible (respectively, infected) degree-k node is infected at time t. The system of equations in Eq. (A.6) consists of at most $3k_{max} + 1$ differential equations, and it typically gives solutions that are more accurate than the MF equations in Eq. (A.4).

Equations (A.4) and (A.6) reduce to known MF and PA approximations for several well-studied dynamical system on networks. For example, consider the SIS disease model. Inserting $F_{k,m} = \lambda m$ and $R_{k,m} = \mu$ from Eqs. (A.1) and (A.2) into these general equations yields the MF equations that were derived in Ref. [241] and the PA equations from Refs. [88, 188]. Similarly, using the voter-model rates from Eq. (A.3) yields the MF equations of Ref. [286]. The voter-model PA equations that one obtains in this way [107] constitute a dynamical system whose dimensionality lies between those of Refs. [260] and [315].

References [107, 108] derived the general MF and PA equations in Eqs. (A.4) and (A.6), respectively, by considering a more complicated approximation scheme that involves what are called *approximate master equations (AME)*. In this AME system, one divides nodes into compartments based both on node state and on the number of infected neighbors. One approximates transitions between compartments by global rates to yield a system of (at most) $[(k_{max} + 1)(k_{max} + 2)]$ closed nonlinear differential equations. (See Sec. III of Ref. [108].) One can then derive PA equations from the AME system by assuming that the number m of infected neighbors of a susceptible (respectively, infected) node is distributed according to the binomial distribution $B_{k,m}(p_k)$ [respectively, $B_{k,m}(q_k)$] and then using this ansatz to derive Eq. (A.6) for $p_k(t)$ and $q_k(t)$. This ansatz is exact when it is correct to treat the neighbors of a node as independent of each other. As we discussed in MF assumption (2) of Sec. 4.3.4, dynamical correlations imply that this independence-of-neighbors assumption is not true in general, so the AME solutions typically are more accurate than PA solutions. The MF equations in Eq. (A.4) result from replacing both p_k and q_k by ω. In this situation, one neglects the dependence between the state of the node itself and the states of its neighbors when the node is updated [see MF assumption (1) of Sec. 4.3.4].

A.3 Monotonic Dynamics and Response Functions

For general binary-state dynamics, an AME system like the one discussed in [108] is high-dimensional and difficult to analyze. However, in the special case of monotonic threshold dynamics (see Chapter 3), the AME system reduces to a coupled set of just two ODEs. This dramatically simplifies analysis. Note that this reduction is exact— i.e., it does not involve any approximation—so the two-dimensional (2D) system is more accurate than naive PA or MF theories. See Fig. 4.1 for an example calculation.

A.3.1 Monotonic Threshold Dynamics

Monotonic dynamics allow only one-way transitions, so $R_{k,m} \equiv 0$ for all k and m. Threshold dynamics occur when the transition rate has the form

$$F_{\mathbf{k},m} = \begin{cases} 0, & \text{if } m < M_{\mathbf{k}} \\ 1, & \text{if } m \geq M_{\mathbf{k}} \end{cases} . \tag{A.8}$$

This reflects deterministic infection (or activation) of a node (once it is chosen for updating) when m equals or exceeds the threshold level $M_{\mathbf{k}}$. We have introduced the vector $\mathbf{k}$ to encode two properties of the nodes: their degree k (a scalar) and their type r. (More generally, the type could be a vector $\mathbf{r}$.) Together, these two properties determine the threshold $M_{\mathbf{k}}$ for the nodes in a network. For example, all nodes of type 1 might have the same threshold M_1, and all nodes of type 2 might have a common threshold M_2. Thus, the set of nodes is partitioned into disjoint sets that are labeled by their degree and their type. The type might be some external label for a node (e.g., the dormitory residence of a student in a university Facebook network [307]), the assignment of a node into some community [213], an indicator of whether a node is immune to peer pressure [268], or something else. We combine the degree and type labels into a 2-vector by writing $\mathbf{k} = (k, r)$ for the $\mathbf{k}$ class of nodes. We generalize the degree distribution to the joint distribution $P_{\mathbf{k}}$, which gives the probability that a node that is chosen uniformly at random has feature vector $\mathbf{k}$ (i.e., it has degree k and is of type r). If the threshold of the nodes are chosen at random using a process that doesn't depend on k (i.e, the thresholds are independent of the node degrees), then one can factor the $P_{\mathbf{k}}$ distribution as $P_{\mathbf{k}} = P_k P_r$, where P_k is the degree distribution and P_r is the probability that a node is of type r.

By considering a large set of discrete types, it is possible to approximate a continuous distribution of thresholds (e.g., a Gaussian distribution) to a desired level of accuracy. Reference [108] demonstrated that the AME system reduces exactly to the pair of ODEs

$$\frac{d\rho}{dt} = h(\phi) - \rho \,, \tag{A.9}$$

$$\frac{d\phi}{dt} = g(\phi) - \phi \,, \tag{A.10}$$

where the functions $h(\phi)$ and $g(\phi)$ are

$$h(\phi) = \rho(0) + [1 - \rho(0)] \sum_{\mathbf{k}} P_{\mathbf{k}} \sum_{m \geq M_{\mathbf{k}}} B_{k,m}(\phi) \,, \tag{A.11}$$

$$g(\phi) = \rho(0) + [1 - \rho(0)] \sum_{\mathbf{k}} \frac{k}{z} P_{\mathbf{k}} \sum_{m \geq M_{\mathbf{k}}} B_{k-1,m}(\phi) \,. \tag{A.12}$$

The variable $\phi(t)$ is the probability that a node at one end of an edge is infected, conditional on the node attached to the other end of the edge being susceptible [106]. The initial conditions for Eqs. (A.9, A.10) are

$$\phi(0) = \rho(0) = \sum_{\mathbf{k}} P_{\mathbf{k}} \rho_{\mathbf{k}}(0) \,. \tag{A.13}$$

Solving (A.9, A.10) with the initial conditions (A.13) yields the fraction $\rho(t)$ of infected nodes in a network at time t to a very high level of accuracy. See Fig. 4.1 for some examples. Note that the AME solutions (red curves) are identical to the solution of the 2D system (A.9, A.10), and they match the numerical simulations of the original system (black symbols) very closely.

A.3.2   Response Functions for Monotonic Binary Dynamics

Although Ref. [108] derived Eqs. (A.9, A.10) from a full AME system, it is also possible to obtain these equations using other approaches. For example, [106] used methods from the study of zero-temperature random-field Ising models [74, 110] to derive (A.9, A.10) as the asynchronous limit of corresponding synchronous-updating equations. These methods are related to belief-propagation and message-passing algorithms [214], and the approach of Ref. [106] makes it possible to generalize beyond the (uncorrelated) configuration-model networks that we have considered thus far. Low-dimensional descriptions such as (A.9, A.10) for monotonic threshold dynamics have also been derived for networks with degree–degree correlations and/or community structure [106, 213], networks with non-negligible clustering [129], and multiplex networks [333]. In Ref. [106], it was also demonstrated that one can express several structural characteristics of networks, such as the sizes of K-cores [79] and the giant connected component (GCC) sizes for site and bond percolation, as steady states of monotonic threshold dynamics. Such characteristics can thereby be determined using equations of the form (A.9, A.10). This perspective is likely to be particularly fruitful in the extension of traditional network measures to multiplex networks and other multilayer networks [28, 49, 173, 333].

As an example, we show how to apply Eqs. (A.9, A.10) to the Watts threshold model for social influence (see Sec. 3.3.1). Recall that each node has a threshold R_i that is drawn from some distribution P_R, and a node i that is being updated becomes active (i.e., moves to the "infected" state) if its fraction of active neighbors $m_i/k_i \geq R_i$. The thresholds R_i are distributed independently of node degrees, so $P_k = P_k P_R$. Equation (A.8) for this model is

$$F_{\mathbf{k},m} = \begin{cases} 0, & \text{if } m < kR \\ 1, & \text{if } m \geq kR \end{cases}, \tag{A.14}$$

and the sums over types in Eq. (A.11) become

$$\sum_{\mathbf{k}} P_{\mathbf{k}} \sum_{m \geq M_{\mathbf{k}}} B_{k,m}(\phi) = \sum_R P_R \sum_{m \geq kR} B_{k,m}(\phi)$$

$$= \sum_{m=0}^{k} B_{k,m}(\phi) \int_{-\infty}^{\infty} P_R H(m - kR)\, dR$$

$$= \sum_{m=0}^{k} B_{k,m}(\phi) \int_{-\infty}^{\frac{m}{k}} P_R\, dR$$

$$= \sum_{m=0}^{k} B_{k,m}(\phi) C\left(\frac{m}{k}\right), \tag{A.15}$$

where H denotes the Heaviside function and C is the cumulative distribution function (CDF) of the thresholds.

Reference [106] showed that the $C(m/k)$ term in Eq. (A.15) is an example of a *response function* for monotonic dynamics. A response function encapsulates the mechanism by which a susceptible node becomes infected when it is updated. For the present discussion, we suppose that a response function $f(k, m)$ depends on the degree k of a node and the number m of infected neighbors. (More generally, a response function can also depend on other network characteristics.) In this setting, Eqs. (A.11, A.12) take the form

$$h(\phi) = \rho(0) + [1 - \rho(0)] \sum_k P_k \sum_{m=0}^{k} B_{k,m}(\phi) f(k, m), \tag{A.16}$$

$$g(\phi) = \rho(0) + [1 - \rho(0)] \sum_k \frac{k}{z} P_k \sum_{m=0}^{k-1} B_{k-1,m}(\phi) f(k, m). \tag{A.17}$$

Response functions of this type have also been defined for other monotonic threshold models (such as many of the complex-contagion models in Chapter 3) and related processes. In bond percolation, for example, the equations for the size of the GCC (i.e., the steady-state epidemic size in an SIR model) relative to the total number of nodes, which were previously derived using generating-function methods [46], are reproduced by taking the limit $\rho(0) \to 0$, using the response function

$$f(k, m) = \begin{cases} 0, & \text{if } m = 0 \\ 1 - (1 - p)^m, & \text{if } m > 0 \end{cases}, \tag{A.18}$$

and applying the identity

$$\sum_{m=1}^{k} B_{k,m}(\phi) \left[1 - (1 - p)^m\right] = 1 - (1 - p\phi)^k. \tag{A.19}$$

See Sec. II.B of [106] for additional discussion.

A.3.3   Cascade Conditions

In addition to giving very accurate predictions for monotonic dynamics, Eqs. (A.9, A.10) make it possible to obtain analytical insights into dynamical processes. Let's again use the Watts threshold model as an example. Consider the question of whether, for a given distribution of thresholds, global cascades can occur on configuration-model networks with degree distribution P_k. Watts used percolation theory in [328] to derive a *cascade condition* that addresses this question. We will now derive the same cascade condition by using linear stability analysis of the 2D dynamical system defined by Eqs. (A.9, A.10, A.16, A.17).

Suppose that all nodes have positive thresholds, so $C(0) = 0$, and consider the seed fraction $\rho(0)$ to be vanishingly small. The system (A.9, A.10) then has an equilibrium point at $(\rho, \phi) = (0, 0)$ that corresponds to a complete absence of infection. However, if this equilibrium point is (linearly) unstable, then a small perturbation (e.g., the infection of a single node) can move the dynamical system away from the equilibrium at $(0,0)$ and result in values $\rho(t) > 0$, constituting a global cascade. By "global," we mean that a nonzero fraction of the nodes in a network become infected in the $N \to \infty$ limit of infinite network size.

The stability of the system in Eqs. (A.9, A.10) is controlled by Eq. (A.10). When this 1D dynamical system is unstable at $\phi = 0$, the values of ρ determined from Eq. (A.9) are also strictly positive. One uses standard linear stability analysis [82, 292] to show that a local instability occurs at $\phi = 0$ if and only if the following condition holds:

$$\frac{d}{d\phi} [g(\phi) - \phi] \bigg|_{\phi=0} > 0 . \tag{A.20}$$

That is, there is a local instability when $g'(0) > 1$, and the monotonic nature of the dynamics guarantees in this case that $\rho(t)$ is strictly positive for all time (i.e., that a global cascade occurs). We now differentiate Eq. (A.17), which is the general response-function form of g, using $f(k,m) = C(m/k)$ for the Watts model and incorporating the facts that $B_{k,m}(0) = \delta_{m,0}$ and $f(k,0) = 0$. This yields the cascade condition that was found in [328]:

$$\sum_k \frac{k}{z}(k-1)P_k f(k,1) > 1 . \tag{A.21}$$

When Eq. (A.21) is satisfied, global cascades can occur in the Watts model. Note that this general condition incorporates information about both network topology (via the degree distribution P_k) and node-level dynamics (through the expected response $f(k,1)$ of a degree-k node to a single infected neighbor). It thereby yields interesting and useful information on how network topology influences dynamics. Further generalizations were considered recently in Ref. [268].

In closing, we note that there are several ways to define a "cascade," especially in the context of more general applications than the specific models that we have considered. When studying empirical data, one may wish to define a "cascade" based on a specified minimum fraction of nodes that eventually become infected or a specified minimum fraction that become infected within a specified amount of time. In practical situations, it can be important to consider the "infection" of people with a meme (or with something else) within a finite duration of time, so the $t \to \infty$ limit that one typically considers to compute cascade sizes is too restrictive on some occasions. For example, looser notions of a cascade are relevant when considering social influence on networks in the commercial and governmental sectors, as it may be necessary to convince as many people as possible to adopt an idea in a rather limited amount of time. One can also use heterogeneous measures by, for example, separately considering whether different parts (e.g., communities) of a network are affected by a contagion to different extents or by measuring the depth of infection chains starting from different nodes. As an example, see the "structural virality" diagnostic of [117].

References

1. J.A. Acebrón, L.L. Bonilla, V. Pérez, F. Ritort, R. Spigler, The Kuramoto model: A simple paradigm for synchronization phenomena. Rev. Mod. Phys. **77**(1), 137–185 (2005)
2. D. Achlioptas, R.M. D'Souza, J. Spencer, Explosive percolation in random networks. Science **323**(5920), 1453–1455 (2009)
3. J. Adler, Bootstrap percolation. Physica A **171**(3), 453–470 (1991)
4. M. Aizenman, K. Lebowitz, Metastability effects in bootstrap percolation. J. Phys. A **21**(19), 3801–3813 (1988)
5. D. Aldous, J.A. Fill, Reversible Markov Chains and Random Walks on Graphs (2002). Unfinished monograph, recompiled 2014. Available at http://www.stat.berkeley.edu/~aldous/RWG/book.html
6. S. Aral, L. Muchnik, A. Sundararajan, Distinguishing influence-based contagion from homophily-driven diffusion in dynamic networks. Proc. Natl. Acad. Sci. U. S. A. **106**(51), 21544–21549 (2009)
7. S. Aral, D. Walker, Tie strength, embeddedness, and social influence: A large-scale networked experiment. Manag. Sci. **60**(6), 1352–1370 (2014)
8. A. Arenas, A. Díaz-Guilera, C.J. Pérez-Vicente, Synchronization processes in complex networks. Physica D **224**(1–2), 27–34 (2006)
9. A. Arenas, A. Díaz-Guilera, C.J. Pérez-Vicente, Synchronization reveals topological scales in complex networks. Phys. Rev. Lett. **96**(11), 114102 (2006)
10. A. Arenas, A. Díaz-Guilera, J. Kurths, Y. Moreno, C. Zhou, Synchronization in complex networks. Phys. Rep. **469**(3), 93–153 (2008)
11. P. Ashwin, S. Coombes, R. Nicks, Mathematical frameworks for oscillatory network dynamics in neuroscience. J. Math. Neurosci. **6**, 2 (2016)
12. M. Asllani, D.M. Busiello, T. Carletti, D. Fanelli, G. Planchon, Turing patterns in multiplex networks. Phys. Rev. E **90**(4), 042814 (2014)
13. R.J. Bagley, J.D. Farmer, S.A. Kauffman, N.H. Packard, A.S. Perelson, I.M. Stadnyk, Modeling adaptive biological systems. Biosystems **23**(2–3), 113–137 (1989)
14. E. Bakshy, Showing support for marriage equality on Facebook (2013). Available at https://www.facebook.com/notes/facebook-data-science/showing-support-for-marriage-equality-on-facebook/10151430548593859
15. D. Balcan, H. Hu, B. Gonçalves, P. Bajardi, C. Poletto, J. Ramasco, D. Paolotti, N. Perra, M. Tizzoni, W. Van den Broeck, V. Colizza, A. Vespignani, Seasonal transmission potential and activity peaks of the new influenza A/H1N1: A Monte Carlo likelihood analysis based on human mobility. BMC Med. **7**(1), 45 (2009)

© Springer International Publishing Switzerland 2016
M.A. Porter, J.P. Gleeson, *Dynamical Systems on Networks*, Frontiers in Applied
Dynamical Systems: Reviews and Tutorials 4, DOI 10.1007/978-3-319-26641-1

16. A.L. Barabási, R. Albert, Emergence of scaling in random networks. Science **286**(5439), 509–512 (1999)
17. M. Barahona, L. Pecora, Synchronization in small world systems. Phys. Rev. Lett. **89**(5), 054101 (2002)
18. A. Barrat, M. Barthelemy, A. Vespignani, *Dynamical Processes on Complex Networks* (Cambridge University Press, Cambridge, 2008)
19. J. Bascompte, P. Jordano, *Mutualistic Networks* (Princeton University Press, Princeton, 2013)
20. F.M. Bass, A new product growth for model consumer durables. Manag. Sci. **15**(5), 215–227 (1969)
21. K.E. Bassler, C.I. Del Genio, P. Erdős, I. Miklós, Z. Toroczkai, Exact sampling of graphs with prescribed degree correlations. New J. Phys. **17**(8), 083052 (2015)
22. F. Battiston, A. Cairoli, V. Nicosia, A. Baule, V. Latora, Interplay between consensus and coherence in a model of interacting opinions (2015). arXiv:1506.04544
23. G.J. Baxter, S.N. Dorogovtsev, A.V. Goltsev, J.F. Mendes, Bootstrap percolation on complex networks. Phys. Rev. E **82**(1), 011103 (2010)
24. V.N. Belykh, I.V. Belykh, M. Hasler, Connection graph stability method for synchronized coupled chaotic systems. Physica D **195**(1–2), 159–187 (2004)
25. I.V. Belykh, V.N. Belykh, M. Hasler, Blinking model and synchronization in small-world networks with a time-varying coupling. Physica D **195**(1–2), 188–206 (2004)
26. C. Bick, M. Field, Asynchronous networks and event driven dynamics (2015). arXiv:1509.04045
27. S. Boccaletti, V. Latora, Y. Moreno, M. Chavez, D.-U. Hwang, Complex networks: Structure and dynamics. Phys. Rep. **424**(4), 175–308 (2006)
28. S. Boccaletti, G. Bianconi, R. Criado, C.I. del Genio, J. Gómez-Gardeñes, M. Romance, I. Sendiña-Nadal, Z. Wang, M. Zanin, The structure and dynamics of multilayer networks. Phys. Rep. **544**(1), 1–122 (2014)
29. M. Boguñá, R. Pastor-Satorras, Epidemic spreading in correlated complex networks. Phys. Rev. E **66**(4), 047104 (2002)
30. M. Boguñá, L.F. Lafuerza, R. Toral, M.A. Serrano, Simulating non-Markovian stochastic processes. Phys. Rev. E **90**(4), 042108 (2014)
31. G.A. Böhme, T. Gross, Analytical calculation of fragmentation transitions in adaptive networks. Phys. Rev. E **83**(3), 35101 (2011)
32. K.A. Bold, K. Rajendran, B. Ráth, I.G. Kevrekidis, An equation-free approach to coarse-graining the dynamics of networks. J. Comput. Dyn. **1**(1), 111–134 (2014)
33. B. Bollobás, *Modern Graph Theory* (Springer, New York, 1998)
34. B. Bollobás, *Random Graphs*, 2nd edn. (Cambridge University Press, Cambridge, 2011)
35. J. Borge-Holthoefer, R.A. Baños, S. González-Bailón, Y. Moreno, Cascading behavior in complex socio-technical networks. J. Complex Networks **1**(1), 3–24 (2013)
36. D. Braess, Über ein paradoxon aus der verkehrsplanung. Unternehmensforschung **12**, 258–268 (1969)
37. F. Brauer, C. Castillo-Chavez, *Mathematical Models in Population Biology and Epidemiology*, 2nd edn. (Springer, New York, 2012)
38. A. Bressan, S. Canić, M. Garavello, M. Herty, B. Piccoli, Flows on networks: Recent results and perspectives. EMS Surv. Math. Sci. **1**(1), 47–111 (2014)
39. S. Brin, L. Page, The anatomy of a large-scale hypertextual Web search engine. Comput. Netw. ISDN Syst. **30**(1–7), 107–117 (1998) (Proceedings of the Seventh International World Wide Web Conference)
40. C. Brownlees, C. Hans, E. Nualart, Bank credit risk networks: Evidence from the Eurozone crises (2014). Available at http://www.greta.it/credit/credit2014/PAPERS/Posters/Thursday/Brownless_Hans_Nualart.pdf
41. N. Brunel, V. Hakim, Fast global oscillations in networks of integrate-and-fire neurons with low firing rates. Neural Comput. **11**(7), 1621–1671 (2002)
42. L.A. Bunimovich, B.Z. Webb, Isospectral compression and other useful isospectral transformations of dynamical networks. Chaos **22**(3), 033118 (2012)

43. L.A. Bunimovich, B.Z. Webb, Isospectral graph transformations, spectral equivalence, and global stability of dynamical networks. Nonlinearity 25(1), 211–254 (2012)
44. F. Caccioli, M. Shrestha, C. Moore, J.D. Farmer, Stability analysis of financial contagion due to overlapping portfolios. J. Bank. Finance 46, 233–245 (2014)
45. C. Cakan, J. Lehnert, E. Schöll, Heterogeneous delays in neural networks. Eur. Phys. J. B 87(3), 54 (2014)
46. D.S. Callaway, M.E.J. Newman, S.H. Strogatz, D.J. Watts, Network robustness and fragility: Percolation on random graphs. Phys. Rev. Lett. 85(25), 5468 (2000)
47. C. Castellano, V. Loreto, A. Barrat, F. Cecconi, D. Parisi, Comparison of voter and Glauber ordering dynamics on networks. Phys. Rev. E 71(6), 066107 (2005)
48. C. Castellano, S. Fortunato, V. Loreto, Statistical physics of social dynamics. Rev. Mod. Phys. 81(2), 591–646 (2009)
49. D. Cellai, E. López, J. Zhou, J.P. Gleeson, G. Bianconi, Percolation in multiplex networks with overlap. Phys. Rev. E 88(5), 052811 (2013)
50. D. Centola, M. Macy, Complex contagions and the weakness of long ties. Am. J. Sociol. 113(3), 702–734 (2007)
51. J. Chalupa, P.L. Leath, G.R. Reich, Bootstrap percolation on a Bethe lattice. J. Phys. C Solid State Phys. 12(1), L31–L35 (1979)
52. M. Chaves, E.D. Sontag, R. Albert, Robustness and fragility of Boolean models for genetic regulatory networks. J. Theor. Biol. 235(3), 431–449 (2005)
53. N.A. Christakis, J.H. Fowler, The spread of obesity in a large social network over 32 years. N. Engl. J. Med. 357(4), 370–379 (2007)
54. N.A. Christakis, J.H. Fowler, Social contagion theory: Examining dynamic social networks and human behavior. Stat. Med. 32(4), 556–577 (2013)
55. A. Clauset, C.R. Shalizi, M.E.J. Newman, Power-law distributions in empirical data. SIAM Rev. 51(4), 661–703 (2009)
56. P. Clifford, A. Sudbury, A model for spatial conflict. Biometrika 60(3), 581–588 (1973)
57. V. Colizza, A. Barrat, M. Barthelemy, A.-J. Valleron, A. Vespignani, Modeling the world-wide spread of pandemic influenza: Baseline case and containment interventions. PLoS Med. 4(1), e13 (2007)
58. V. Colizza, R. Pastor-Satorras, A. Vespignani, Reaction-diffusion processes and metapopulation models in heterogeneous networks. Nat. Phys. 3(4), 276–282 (2007)
59. S. Contemori, F.D. Patti, D. Fanelli, F. Miele, Multiple scale theory of topology driven pattern on directed networks (2015). arXiv:1508.00148
60. N.J. Cowan, E.J. Chastain, D.A. Vilhena, J.S. Freudenberg, C.T. Bergstrom, Nodal dynamics, not degree distributions, determine the structural controllability of complex networks. PLoS One 7(6), e38398 (2012)
61. G. Craciun, M. Feinberg, Multiple equilibria in complex chemical reaction networks. I. The injectivity property. SIAM J. Appl. Math. 65(5), 1526–1546 (2005)
62. G. Craciun, M. Feinberg, Multiple equilibria in complex chemical reaction networks. II. The species-reaction graph. SIAM J. Appl. Math. 66(4), 1321–1338 (2006)
63. M.C. Cross, P.C. Hohenberg, Pattern formation outside equilibrium. Rev. Mod. Phys. 65(3), 851–1112 (1993)
64. P. Csermely, A. London, L.-Y. Wu, B. Uzzi, Structure and dynamics of core–periphery networks. J. Complex Networks 1(2), 93–123 (2013)
65. P. Cvitanović, R. Artuso, R. Mainieri, G. Tanner, G. Vattay, Chaos: Classical and Quantum, (Niels Bohr Institute, Copenhagen, 2015). Available at http://www.chaosbook.org
66. R.A. da Costa, S.N. Dorogovstev, A.V. Goltsev, J.F.F. Mendes, Explosive percolation transition is actually continuous. Phys. Rev. Lett. 105(25), 255701 (2010)
67. M.J. de Oliveira, Isotropic majority-vote model on a square lattice. J. Stat. Phys. 66(1–2), 273–281 (1992)
68. D.J. de Solla Price, A general theory of bibliometric and other cumulative advantage processes. J. Am. Soc. Inf. Sci. 27(5), 292–306 (1976)
69. M.H. DeGroot, Reaching consensus. J. Am. Stat. Assoc. 69(345), 118–121 (1974)

70. P. DeLellis, M. di Bernardo, T.E. Gorochowski, G. Russo, Synchronization and control of complex networks via contraction, adaptation and evolution. IEEE Circuits Syst. Mag. **10**(3), 64–82 (third quarter) (2010)

71. G. Demirel, F. Vázquez, G.A. Bhöme, T. Gross, Moment-closure approximations for discrete adaptive networks. Physica D **267**(1), 68–80 (2014)

72. B.A. Desmarais, S.J. Cranmer, Statistical mechanics of networks: Estimation and uncertainty. Physica A **391**(4), 1865–1876 (2012)

73. M. Dhamala, V.K. Jirsa, M. Ding, Enhancement of neural synchrony by time delay. Phys. Rev. Lett. **92**(7), 074104 (2004)

74. D. Dhar, P. Shukla, J.P. Sethna, Zero-temperature hysteresis in the random-field Ising model on a Bethe lattice. J. Phys. A Math. Gen. **30**(15), 5259 (1997)

75. A.-L. Do, S. Boccaletti, T. Gross, Graphical notation reveals topological stability criteria for collective dynamics in complex networks. Phys. Rev. Lett. **108**(19), 194102 (2012)

76. P.S. Dodds, K.D. Harris, C.M. Danforth, Limited imitation contagion on random networks: Chaos, universality, and unpredictability. Phys. Rev. Lett. **110**(15), 158701 (2013)

77. P.S. Dodds, D.J. Watts, Universal behavior in a generalized model of contagion. Phys. Rev. Lett. **92**(21), 218701 (2004)

78. P.S. Dodds, D.J. Watts, A generalized model of social and biological contagion. J. Theor. Biol. **232**(4), 587–604 (2005)

79. S.N. Dorogovtsev, A.V. Goltsev, J.F.F. Mendes, K-core organization of complex networks. Phys. Rev. Lett. **96**(4), 040601 (2006)

80. S.N. Dorogovtsev, A.V. Goltsev, J.F.F. Mendes, Critical phenomena in complex networks. Rev. Mod. Phys. **80**(4), 1275–1335 (2008)

81. D.M. Dozier, Communication Networks and the Role of Thresholds in the Adoption of Innovations. Ph.D. Thesis, Stanford University (1977)

82. P.G. Drazin, *Nonlinear Systems* (Cambridge University Press, Cambridge, 1992)

83. J. Drury, 2011—The year of contagion? (blog entry in *The Crowd*; 1 Jan. 2012) Available at http://drury-sussex-the-crowd.blogspot.co.uk/2012/01/2011-year-of-contagion.html

84. R.M. D'Souza, J. Nagler, Anomalous critical and supercritical phenomena in explosive percolation. Nat. Phys. **11**(7), 531–538 (2015)

85. R. Durrett, *Random Graph Dynamics* (Cambridge University Press, Cambridge, 2007)

86. R. Durrett, Some features of the spread of epidemics and information on a random graph. Proc. Natl. Acad. Sci. U. S. A. **107**(10), 4491–4498 (2010)

87. R. Durrett, J.P. Gleeson, A.L. Lloyd, P.J. Mucha, F. Shi, D. Sivakoff, J.E. Socolar, C. Varghese, Graph fission in an evolving voter model. Proc. Natl. Acad. Sci. U. S. A. **109**(10), 3682–3687 (2012)

88. K.T.D. Eames, M.J. Keeling, Modeling dynamic and network heterogeneities in the spread of sexually transmitted diseases. Proc. Natl. Acad. Sci. U. S. A. **99**(20), 13330–13335 (2002)

89. D. Easley, J. Kleinberg, *Networks, Crowds, and Markets: Reasoning About a Highly Connected World* (Cambridge University Press, Cambridge, 2010)

90. J. Epperlain, A.-L. Do, T. Gross, S. Siegmund, Meso-scale obstructions to stability of 1D center manifolds for networks of coupled differential equations with symmetric Jacobian. Physica D **261**(3), 1–7 (2013)

91. G.B. Ermentrout, D.H. Terman, *Mathematical Foundations of Neuroscience* (Springer, Berlin, 2010)

92. J.D. Farmer, S.A. Kauffman, N.H. Packard, Autocatalytic replication of polymers. Physica D **22**(1), 50–67 (1986)

93. M. Feinberg, Lectures on Chemical Reaction Networks. Mathematics Research Center, University of Wisconsin (1979). Available at http://crnt.engineering.osu.edu/LecturesOnReactionNetworks

94. J. Fernández-Gracia, K. Suchecki, J.J. Ramasco, M. San Miguel, V.M. Eguíluz, Is the voter model a model for voters? Phys. Rev. Lett. **112**(15), 158701 (2014)

95. P.G. Fennell, S. Melnik, J.P. Gleeson, The limitations of discrete-time approaches to continuous-time contagion dynamics (2016). arXiv:1603.01132
96. K.S. Fink, G. Johnson, T. Carroll, D. Mar, L. Pecora, Three coupled oscillators as a universal probe of synchronization stability in coupled oscillator arrays. Phys. Rev. E **61**(5), 5080–5090 (2000)
97. R.A. Fisher, *The Genetical Theory of Natural Selection*, Complete Varorium Edition (Oxford University Press, Oxford, 1999)
98. N.E. Friedkin, E.C. Johnsen, *Social Influence Network Theory* (Cambridge University Press, Cambridge, 2011)
99. X. Fu, M. Small, G. Chen, *Propagation Dynamics on Complex Networks: Models, Methods and Stability Analysis* (Wiley, New York, 2014)
100. S. Funk, M. Salathé, V.A.A. Jansen, Modelling the influence of human behaviour on the spread of infectious diseases: A review. J. R. Soc. Interface **7**(50), 1247–1256 (2010)
101. P.M. Gade, C.-K. Hu, Synchronous chaos in coupled map lattices with small-world interactions. Phys. Rev. E **62**(5), 6409–6413 (2000)
102. P. Gai, S. Kapadia, Contagion in financial networks. Proc. R. Soc. A **466**(2120), 2401–2423 (2010)
103. M. Garavello, B. Piccoli, *Traffic Flow on Networks* (American Institute of Mathematical Sciences, San Jose, 2008)
104. T. Gedeon, S. Harker, H. Kokubu, K. Mischaikow, H. Ok, Global dynamics for steep sigmoidal nonlinearities in two dimensions (2015). arXiv:1508.02438
105. D.T. Gillespie, Exact stochastic simulation of coupled chemical reactions. J. Phys. Chem. **81**(25), 2340–2361 (1977)
106. J.P. Gleeson, Cascades on correlated and modular random networks. Phys. Rev. E **77**(4), 046117 (2008)
107. J.P. Gleeson, High-accuracy approximation of binary-state dynamics on networks. Phys. Rev. Lett. **107**(6), 068701 (2011)
108. J.P. Gleeson, Binary-state dynamics on complex networks: Pair approximation and beyond. Phys. Rev. X **3**(2), 021004 (2013)
109. J.P. Gleeson, D.J. Cahalane, An analytical approach to cascades on random networks, in *SPIE Fourth International Symposium on Fluctuations and Noise* (International Society for Optics and Photonics, 2007), 66010W
110. J.P. Gleeson, D.J. Cahalane, Seed size strongly affects cascades on random networks. Phys. Rev. E **75**(5), 056103 (2007)
111. J.P. Gleeson, K.P. O'Sullivan, R.A. Baños, Y. Moreno, Determinants of meme popularity (2015). arXiv:1501.05956
112. J.P. Gleeson, S. Melnik, J.A. Ward, M.A. Porter, P.J. Mucha, Accuracy of mean-field theory for dynamics on real-world networks. Phys. Rev. E **85**(2), 026106 (2012)
113. J.P. Gleeson, J.A. Ward, K.P. O'Sullivan, W.T. Lee, Competition-induced criticality in a model of meme popularity. Phys. Rev. Lett. **112**(4), 048701 (2014)
114. D.F. Gleich, PageRank beyond the Web. SIAM Rev. **57**(3), 321–363 (2015)
115. S. Gnutzmann, U. Smilansky, Quantum graphs: Applications to quantum chaos and universal spectral statistics. Adv. Phys. **55**(5–6), 527–625 (2006)
116. S. Gnutzmann, D. Waltner, Stationary waves on nonlinear quantum graphs I: General framework and canonical perturbation theory (2015). arXiv:1510.00351
117. S. Goel, A. Anderson, J. Hofman, D.J. Watts, The structural virality of online diffusion. Manag. Sci. **62**(1), 180–196 (2016).
118. M. Golubitsky, R. Lauterbach, Bifurcations from synchrony in homogeneous networks: Linear theory. SIAM J. Appl. Dyn. Syst. **8**(1), 40–75 (2009)
119. M. Golubitsky, I. Stewart, Patterns of synchrony in coupled cell networks with multiple arrows. SIAM J. Appl. Dyn. Syst. **4**(1), 78–100 (2005)
120. J. Gómez-Gardeñes, Y. Moreno. From scale-free to Erdős–Rényi networks. Phys. Rev. E **73**(5), 056124 (2006)
121. J. Gómez-Gardeñes, S. Gómez, A. Arenas, Y. Moreno, Explosive synchronization transitions in scale-free networks. Phys. Rev. Lett. **106**(12), 128701 (2011)

122. M. Granovetter, Threshold models of collective behavior. Am. J. Sociol. **83**(6), 1420–1443 (1978)
123. P. Grassberger, On the critical behavior of the general epidemic process and dynamical percolation. Math. Biosci. **63**(2), 157–172 (1983)
124. T. Gross, B. Blasius, Adaptive coevolutionary networks: A review. J. R. Soc. Interface **5**(20), 259–271 (2008)
125. T. Gross, C.J. Dommar D'Lima, B. Blasius, Epidemic dynamics on an adaptive network. Phys. Rev. Lett. **96**(20), 208701 (2006)
126. E. Gross, H.A. Harrington, Z. Rosen, B. Sturmfels, Algebraic systems biology: A case study for the Wnt pathway (2015). arXiv:1502.03188
127. J. Guckenheimer, P. Holmes, *Nonlinear Oscillations, Dynamical Systems, and Bifurcations of Vector Fields*. Number 42 in Applied Mathematical Sciences (Springer, New York, 1983)
128. S. Gupta, A. Campa, S. Ruffo, Kuramoto model of synchronization: Equilibrium and nonequilibrium aspects. J. Stat. Mech. Theory Exp. **2014**(8), R08001 (2014)
129. A. Hackett, J.P. Gleeson, Cascades on clique-based graphs. Phys. Rev. E **87**(6), 062801 (2013)
130. A.G. Haldane, R.M. May, Systemic risk in banking ecosystems. Nature **469**(7330), 351–355 (2011)
131. H.A. Harrington, K.L. Ho, T. Thorne, M.P.H. Stumpf, Parameter-free model discrimination criterion based on steady-state coplanarity. Proc. Natl. Acad. Sci. U. S. A. **109**(39), 15746–15751 (2012)
132. L. Hébert-Dufresne, O. Patterson-Lomba, G.M. Goerg, B.M. Althouse, Pathogen mutation modeled by competition between site and bond percolation. Phys. Rev. Lett. **110**(10), 108103 (2013)
133. H.W. Hethcote, The mathematics of infectious diseases. SIAM Rev. **42**(4), 599–653 (2000)
134. P.D.H. Hines, I. Dobson, P. Rezaei, Cascading power outages propagate locally in an influence graph that is not the actual grid topology (2015). arXiv:1508.01775
135. T. Hoffmann, M.A. Porter, R. Lambiotte, Generalized master equations for non-Poisson dynamics on networks. Phys. Rev. E **86**(4), 046102 (2012)
136. T. Hoffmann, M.A. Porter, R. Lambiotte, Random walks on stochastic temporal networks, in *Temporal Networks* (Springer, New York, 2013), pp. 295–314
137. R. Holley, T.M. Liggett, Ergodic theorems for weakly interacting infinite systems and the voter model. Ann. Probab. **3**(4), 643–663 (1975)
138. P. Holme, Model versions and fast algorithms for network epidemiology. J. Logistical Eng. Univ. **30**(3), 1–7 (2014)
139. P. Holme, Modern temporal network theory: A colloquium. Eur. Phys. J. B **88**(9), 234 (2015)
140. P. Holme, M.E.J. Newman, Nonequilibrium phase transition in the coevolution of networks and opinions. Phys. Rev. E **74**(5), 056108 (2006)
141. P. Holme, J. Saramäki, Temporal networks. Phys. Rep. **519**(3), 97–125 (2012)
142. P. Holme, J. Saramäki (eds.), *Temporal Networks* (Springer, New York, 2013)
143. D.X. Horváth, J. Kertész, Spreading dynamics on networks: The role of burstiness, topology and non-stationarity. New J. Phys. **16**(7), 073037 (2014)
144. T. House, Algebraic moment closure for population dynamics on discrete structures. Bull. Math. Biol. **77**(4), 646–659 (2015)
145. T.R. Hurd, J.P. Gleeson, A framework for analyzing contagion in banking networks (2011). arXiv:1110.4312
146. T.R. Hurd, J.P. Gleeson, On Watts' cascade model with random link weights. J. Complex Networks **1**(1), 25–43 (2013)
147. T. Ichinomiya, Frequency synchronization in a random oscillator network. Phys. Rev. E **70**(2), 026116 (2004)
148. T. Ichinomiya, Path-integral approach to dynamics in a sparse random network. Phys. Rev. E **72**(1), 016109 (2005)
149. Y. Ide, H. Izuhara, T. Machida, Turing instability in reaction–diffusion models on complex networks (2014). arXiv:1405.0642

150. J. Ito, K. Kaneko, Spontaneous structure formation in a network of chaotic units with variable connection strengths. Phys. Rev. Lett. **88**(2), 028701 (2002)
151. Y. Itoh, C. Mallows, L. Shepp, Explicit sufficient invariants for an interacting particle system. J. Appl. Probab. **35**(3), 633–641 (1998)
152. Y. Iwamasa, N. Masuda, Networks maximizing the consensus time of voter models. Phys. Rev. E **90**(1), 012816 (2014)
153. M.O. Jackson, *Social and Economic Networks* (Princeton University Press, Princeton, 2010)
154. M.O. Jackson, D. López-Pintado, Diffusion and contagion in networks with heterogeneous agents and homophily. Netw. Sci. **1**(1), 49–67 (2013)
155. M.O. Jackson, L. Yariv, Diffusion, Strategic Interaction, and Social Structure, in *Handbook of Social Economics*, ed. by J. Benhabib, A. Bisin, M.O. Jackson (North Holland Press, Amsterdam, 2011), pp. 646–678
156. M.O. Jackson, Y. Zenou, Games on networks, in *Handbook of Game Theory Vol. 4*, ed. by P. Young, S. Zamir (Elsevier, New York, 2014), pp. 95–163
157. A. Jadbabaie, Flocking in networked systems, in *Encyclopedia of Systems and Control* (Springer, New York, 2015)
158. A. Jadbabaie, J. Lin, A.S. Morse, Coordination of groups of mobile autonomous agents using nearest neighbor rules. IEEE Trans. Autom. Control **48**(6), 988–1001 (2003)
159. A. Jadbabaie, N. Motee, M. Barahona, On the stability of the Kuramoto model of coupled nonlinear oscillators, in *Proceedings of the 2004 American Control Conference*, vol. 5, pp. 4296–4301 (2004)
160. L.G.S. Jeub, P. Balachandran, M.A. Porter, P.J. Mucha, M.W. Mahoney, Think locally, act locally: Detection of small, medium-sized, and large communities in large networks. Phys. Rev. E **91**(1), 012821 (2015)
161. P. Jia, A. Mirtabatabaei, N.E. Friedkin, F. Bullo. Opinion dynamics and the evolution of social power in influence networks, SIAM Rev. **57**(3), 367–397 (2015)
162. H.-H. Jo, J.I. Perotti, K. Kaski, J. Kertész, Analytically solvable model of spreading dynamics with non-Poissonian processes. Phys. Rev. X **4**(1), 011041 (2014)
163. B. Joshi, A. Shiu, A survey of methods for deciding whether a reaction network is multistationary. Math. Model. Nat. Phenom. **10**(5), 47–67 (2015)
164. F. Karimi, P. Holme, Threshold model of cascades in empirical temporal networks. Physica A **392**(16), 3476–3483 (2013)
165. M. Karsai, G. Iñiguez, R. Kikas, K. Kaski, J. Kertész, Local cascades induced global contagion: How heterogeneous thresholds, exogenous effects, and unconcerned behaviour govern online adoption spreading (2016). arXiv:1601.07995
166. M. Karsai, M. Kivelä, R.K. Pan, K. Kaski, J. Kerész, A.-L. Barabási, J. Saramäki, Small but slow world: How network topology and burstiness slow down spreading. Phys. Rev. E **83**(2), 025102(R) (2011)
167. M. Karsai, N. Perra, A. Vespignani, Time-varying networks and the weakness of strong ties. Sci. Rep. **4**, 4001 (2014)
168. S.A. Kauffman, Metabolic stability and epigenesis in randomly constructed genetic nets. J. Theor. Biol. **22**(3), 437–467 (1969)
169. D. Kempe, J. Kleinberg, E. Tardos, Maximizing the spread of influence through a social network, in *Proceedings of the Ninth ACM SIGKDD International Conference on Knowledge Discovery and Data Mining, KDD '03* (ACM, New York, 2003), pp. 137–146
170. E. Kenah, J.M. Robins, Second look at the spread of epidemics on networks. Phys. Rev. E **76**(3), 036113 (2007)
171. H. Kesten, What is . . . percolation? Not. Am. Math. Soc. **53**(5), 572–573 (2006)
172. I.Z. Kiss, G. Röst, Z. Vizi, Generalization of pairwise models to non-Markovian epidemics on networks. Phys. Rev. Lett. **115**(7), 078701 (2015)
173. M. Kivelä, A. Arenas, M. Barthelemy, J.P. Gleeson, Y. Moreno, M.A. Porter. Multilayer networks. J. Complex Networks **2**(3), 203–271 (2014)

174. E.D. Kolaczyk, *Statistical Analysis of Network Data: Methods and Notes* (Springer, New York, 2009)
175. T.G. Kolda, B.W. Bader, Tensor decompositions and applications. SIAM Rev. **51**(3), 455–500 (2009)
176. P.L. Krapivsky, S. Redner, D. Volovik, Reinforcement-driven spread of innovations and fads. J. Stat. Mech. Theory Exp. **2011**(12), P12003 (2011)
177. C. Kuehn, Moment closure — A brief review (2015). arXiv:1505.02190,
178. Y. Kuramoto, *Chemical Oscillations, Waves, and Turbulence* (Dover Press, New York, 1984)
179. Y.M. Lai, M.A. Porter, Noise-induced synchronization, desynchronization, and clustering in globally coupled nonidentical oscillators. Phys. Rev. E **88**(1), 012905 (2013)
180. R. Lambiotte, S. Redner, Dynamics of vacillating voters. J. Stat. Mech. Theory Exp. **2007**(10), L10001 (2007)
181. R. Lambiotte, J.-C. Delvenne, M. Barahona, Random walks, Markov processes and the multiscale modular organization of complex networks. Trans. Netw. Sci. Eng. **1**(2), 76–90 (2015) (see also the precursor of this paper at arXiv:0812.1770, 2008)
182. R. Lambiotte, V. Salnikov, M. Rosvall, Effect of memory on the dynamics of random walks on networks. J. Complex Networks **3**(2), 177–188 (2015)
183. N. Lanchier, The Axelrod model for the dissemination of culture revisited. Ann. Appl. Probab. **22**(2), 860–880 (2012)
184. D.-S. Lee, Synchronization transition in scale-free networks: Clusters of synchrony. Phys. Rev. E **72**(2), 026208 (2005)
185. S.H. Lee, P. Holme, Exploring maps with greedy navigators. Phys. Rev. Lett. **108**(12), 128701 (2012)
186. K.-M. Lee, B. Min, K.-I. Goh, Towards real-world complexity: An introduction to multiplex networks. Eur. Phys. J. B **88**(2), 48 (2015)
187. K. Lerman, X. Yan, X.-Z. Wu, The "majority illusion" in social networks (2015). arXiv:1506.03022
188. S.A. Levin, R. Durrett, From individuals to epidemics. Philos. Trans. R. Soc. Lond. Ser. B Biol. Sci. **351**(1347), 1615–1621 (1996)
189. J. Li, W.-X. Wang, Y.-C. Lai, C. Grebogi, Reconstructing complex networks with binary-state dynamics (2015). arXiv:1511.06852
190. T.M. Liggett, *Interacting Particle Systems* (Springer, New York, 2005)
191. J. Lindquist, J. Ma, P. van den Driessche, F.H. Willeboordse, Effective degree network disease models. J. Math. Biol. **62**(2), 143–164 (2011)
192. Y.-Y. Liu, A.-L. Barabási, Control principles of complex networks (2015). arXiv:1508.05384
193. Y.-Y. Liu, J.-J. Slotine, A.-L. Barabási, Controllability of complex networks. Nature **473**(7346), 167–173 (2011)
194. S. Liu, N. Perra, M. Karsai, A. Vespignani, Controlling contagion processes in activity driven networks. Phys. Rev. Lett. **112**(11), 118702 (2014)
195. E. López, R. Parshani, R. Cohen, S. Carmi, S. Havlin, Limited path percolation in complex networks. Phys. Rev. Lett. **99**(18), 188701 (2007)
196. D. Lusher, J. Koskinen, G. Robins, *Exponential Random Graph Models for Social Networks* (Cambridge University Press, Cambridge, 2013)
197. R. Lyons, The spread of evidence-poor medicine via flawed social-network analysis. Stat. Polit. Pol. **2**(1), 2 (2011)
198. R.S. MacKay, J.-A. Sepulchre, Multistability in networks of weakly coupled bistable units. Physica D **82**(3), 243–254 (1995)
199. A.L. MacLean, Z. Rosen, H. Byrne, H.A. Harrington, Parameter-free methods distinguish Wnt pathway models and guide design of experiments. Proc. Natl. Acad. Sci. U.S.A. **112**(9), 2652–2657 (2015)
200. N. Malik, P.J. Mucha, Role of social environment and social clustering in spread of opinions in coevolving networks. Chaos **23**(4), 043123 (2013)
201. V. Marceau, P.-A. Noël, L. Hébert-Dufresne, A. Allard, L.J. Dubé, Adaptive networks: Coevolution of disease and topology. Phys. Rev. E **82**(3), 036116 (2010)

202. S.A. Marvel, T. Martin, C.R. Doering, D. Lusseau, M.E.J. Newman, The small-world effect is a modern phenomenon (2013). arXiv:1310.2636
203. N. Masuda, K. Klemm, V.M. Eguíluz, Temporal networks: Slowing down diffusion by long lasting interactions. Phys. Rev. Lett. **111**(18), 188701 (2013)
204. N. Masuda, L.E.C. Rocha, A Gillespie algorithm for non-Markovian stochastic processes: Laplace transform approach (2016). arXiv:1601.01490
205. Mathematical and Theoretical Biological Institute, Technical reports archive (2016). Available at http://mtbi.asu.edu/research/archive
206. J.N. Matias, Were all those rainbow profile photos another Facebook study? (28 June 2015) Available at http://www.theatlantic.com/technology/archive/2015/06/were-all-those-rainbow-profile-photos-another-facebook-experiment/397088/
207. N. McCullen, A. Rucklidge, C. Bale, T. Foxon, W. Gale, Multiparameter models of innovation diffusion on complex networks. SIAM J. Appl. Dyn. Syst. **12**(1), 515–532 (2013)
208. G.S. Medvedev, The nonlinear heat equation on dense graphs and graph limits. SIAM J. Math. Anal. **46**(4), 2743–2766 (2014)
209. G.S. Medvedev, X. Tang, Stability of twisted states in the Kuramoto model on Cayley and random graphs. J. Nonlinear Sci. **25**(6), 1169–1208 (2015)
210. D. Mehta, N. Daleo, F. Dörfler, J.D. Hauenstein, Algebraic geometrization of the Kuramoto model: Equilibria and stability analysis (2014). arXiv:1412.0666
211. A. Mellor, M. Mobilia, S. Redner, A.M. Rucklidge, J.A. Ward, Role of Luddism on innovation diffusion. Phys. Rev. E **92**(1), 012806 (2015)
212. S. Melnik, J.A. Ward, J.P. Gleeson, M.A. Porter, Multi-stage complex contagions. Chaos **23**(1), 013124 (2013)
213. S. Melnik, M.A. Porter, P.J. Mucha, J.P. Gleeson, Dynamics on modular networks with heterogeneous correlations. Chaos **24**(2), 023106 (2014)
214. M. Mezard, A. Montanari, *Information, Physics, and Computation* (Oxford University Press, Oxford, 2009)
215. J.C. Miller, A note on a paper by Erik Volz: SIR dynamics in random networks. J. Math. Biol. **62**(3), 349–358 (2011)
216. J.C. Miller, Complex contagions and hybrid phase transitions. J. Complex Networks (2015). doi: 10.1093/comnet/cnv021
217. J.C. Miller, Percolating under one roof (2015). arXiv:1505.01396
218. J.C. Miller, I.Z. Kiss, Epidemic spread in networks: Existing methods and current challenges. Math. Modell. Nat. Phenom. **9**(2), 4–42, 1 (2014)
219. J.C. Miller, E.M. Volz, Model hierarchies in edge-based compartmental modeling for infectious disease spread. J. Math. Biol. **67**(4), 869–899 (2013)
220. D. Mollison, Spatial contact models for ecological and epidemic spread. J. R. Stat. Soc. Ser. B Methodol. **39**(3), 283–326 (1977)
221. E. Mones, N.A.M. Araújo, T. Vicsek, H.J. Herrmann, Shock waves on complex networks. Sci. Rep. **4**, 4949 (2014)
222. A. Montanari, A. Saberi, The spread of innovations in social networks. Proc. Natl. Acad. Sci. U. S. A. **107**(47), 20196–20201 (2010)
223. L. Moreau, Stability of multiagent systems with time-dependent communication links. IEEE Trans. Autom. Control **50**(2), 169–182 (2005)
224. S. Morita, Six susceptible–infected–susceptible models on scale-free networks (2015). arXiv:1508.04451
225. A.E. Motter, Networkcontrology. Chaos **25**(9), 096621 (2015)
226. A.E. Motter, R. Albert, Networks in motion. Phys. Today **65**(4), 43–48 (2012)
227. M.E.J. Newman, The structure and function of complex networks. SIAM Rev. **45**(2), 167–256 (2003)
228. M.E.J. Newman, *Networks: An Introduction* (Oxford University Press, Oxford, 2010)
229. T. Nishikawa, A.E. Motter, Network synchronization landscape reveals compensatory structures, quantization, and the positive effect of negative interactions. Proc. Natl. Acad. Sci. U. S. A. **107**(23), 10342–10347 (2010)

230. T. Nishikawa, A.E. Motter, Y.-C. Lai, F.C. Hoppensteadt, Heterogeneity in oscillator networks: Are smaller worlds easier to synchronize? Phys. Rev. Lett. **91**(1), 014101 (2003)
231. P.-A. Noël, B. Davoudi, R.C. Brunham, L.J. Dubé, B. Pourbohloul, Time evolution of epidemic disease on finite and infinite networks. Phys. Rev. E **79**(2), 026101 (2009)
232. P.-A. Noël, A. Allard, L. Hébert-Dufresne, V. Marceau, L.J. Dubé, Propagation on networks: An exact alternative perspective. Phys. Rev. E **85**(3), 031118 (2012)
233. P.-A. Noël, C.D. Brummitt, R.M. D'Souza, Controlling self-organizing dynamics on networks using models that self-organize. Phys. Rev. Lett. **111**(7), 078701 (2013)
234. C. Nowzari, V.M. Preciado, G.J. Pappas, Analysis and control of epidemics: A survey of spreading processes on complex networks. IEEE Control Syst. Mag. **36**(1), 26–46 (2016)
235. M. Ogura, V.M. Preciado, Stability of spreading processes over time-varying large-scale networks. IEEE Trans. Netw. Sci. Eng. **3**(1), 44–57 (2016)
236. R. Olfati-Saber, Flocking for multi-agent dynamic systems: Algorithms and theory. IEEE Trans. Autom. Control **51**(3), 401–420 (2006)
237. P. Oliver, G. Marwell, R. Teixeira, A theory of the critical mass. I. Interdependence, group heterogeneity, and the production of collective action. Am. J. Sociol. **91**(3), 522–556 (1985)
238. G. Palla, I. Derényi, I. Farkas, T. Vicsek, Uncovering the overlapping community structure of complex networks in nature and society. Nature **435**(7043), 814–818 (2005)
239. M.J. Panaggio, D.M. Abrams, Chimera states: Coexistence of coherence and incoherence in networks of coupled oscillators. Nonlinearity **28**(3), R67–R87 (2015)
240. M. Pascual, J.A. Dunne (eds.), *Ecological Networks: Linking Structure to Dynamics in Food Webs* (Oxford University Press, Oxford, 2006)
241. R. Pastor-Satorras, A. Vespignani, Epidemic spreading in scale-free networks. Phys. Rev. Lett. **86**(14), 3200 (2001)
242. R. Pastor-Satorras, C. Castellano, P. Van Mieghem, A. Vespignani, Epidemic processes in complex networks. Rev. Mod. Phys. **87**(4), 925–979 (2015)
243. L.M. Pecora, T.L. Carroll, Master stability functions for synchronized coupled systems. Phys. Rev. Lett. **80**(10), 2109–2112 (1998)
244. L.M. Pecora, T.L. Carroll, Master stability function for globally synchronized systems, in *Encyclopedia of Computational Neuroscience*, ed. by D. Jaeger, R. Jung (Springer, New York, 2014), pp. 1–13
245. L.M. Pecora, T.L. Caroll, Synchronization of chaotic systems. Chaos **25**(9), 097611 (2015)
246. L.M. Pecora, F. Sorrentino, A.M. Hagerstrom, T.E. Murphy, R. Roy, Cluster synchronization and isolated desynchronization in complex networks with symmetries. Nat. Commun. **5**, 4079 (2014)
247. L. Pellis, T. House, M.J. Keeling, Exact and approximate moment closures for non-Markovian network epidemics. J. Theor. Biol. **382**, 160–177 (2015)
248. M. Perc, J. Gómez-Gardeñes, A. Szolnoki, L.M. Floría, Y. Moreno, Evolutionary dynamics of group interactions on structured populations: A review. J. R. Soc. Interface **10**(80), 20120997 (2013)
249. T. Pereira, S. van Strien, J.S.W. Lamb, Dynamics of coupled maps in heterogeneous random networks (2013). arXiv:1308.5526
250. F.J. Pérez-Reche, J.L. Ludlam, S.N. Taraskin, C.A. Gilligan, Synergy in spreading processes: From exploitative to explorative foraging strategies. Phys. Rev. Lett. **106**(21), 218701 (2011)
251. N. Perra, B. Gonçalves, R. Pastor-Satorras, A. Vespignani, Activity driven modeling of time varying networks. Sci. Rep. **2**, 469 (2012)
252. N. Perra, A. Baronchelli, D. Mocanu, B. Gonçalves, R. Pastor-Satorras, A. Vespignani, Random walks and search in time-varying networks. Phys. Rev. Lett. **109**(23), 238701 (2012)
253. R. Pfitzner, I. Scholtes, A. Garas, C.J. Tessone, F. Schweitzer, Betweenness preference: Quantifying correlations in the topological dynamics of temporal networks. Phys. Rev. Lett. **110**(19), 198701 (2013)
254. P. Piedrahíta, J. Borge-Holthoefer, Y. Moreno, A. Arenas, Modeling self-sustained activity cascades in socio-technical networks. Europhys. Lett. **104**(4), 48004 (2013)
255. A. Pikovsky, M. Rosenblum, Synchronization. Scholarpedia **2**(12), 1459 (2007)

256. A. Pikovsky, M. Rosenblum, J. Kurths, *Synchronization: A Universal Concept in Nonlinear Sciences* (Cambridge University Press, Cambridge, 2003)
257. A. Pomerance, E. Ott, M. Girvan, W. Losert, The effect of network topology on the stability of discrete state models of genetic control. Proc. Natl. Acad. Sci. U. S. A. **106**(20), 8209–8214 (2009)
258. M.A. Porter, Small-world network. Scholarpedia **7**(2), 1739 (2012)
259. M.A. Porter, J.-P. Onnela, P.J. Mucha, Communities in networks. Not. Am. Math. Soc. **56**(9), 1082–1097 1164–1166 (2009)
260. E. Pugliese, C. Castellano, Heterogeneous pair approximation for voter models on networks. Europhys. Lett. **88**(5), 58004 (2009)
261. G.X. Qi, H.B. Huang, C.K. Shen, H.J. Wang, L. Chen, Predicting the synchronization time in coupled-map networks. Phys. Rev. E **77**(5), 056205 (2008)
262. K. Rajendran, I.G. Kevrekidis, Coarse graining the dynamics of heterogeneous oscillators in networks with spectral gaps. Phys. Rev. E **84**(3), 036708 (2011)
263. J.G. Restrepo, E. Ott, B.R. Hunt, Onset of synchronization in large networks of coupled oscillators. Phys. Rev. E **71**(3), 036151 (2005)
264. L.E.C. Rocha, N. Masuda, Individual-based approach to epidemic processes on arbitrary dynamic contact networks (2015). arXiv:1510.09179
265. F.A. Rodrigues, T.K.DM. Peron, P. Ji, J. Kurths, The Kuramoto model in complex networks. Phys. Rep. **610**, 1–98 (2016)
266. E.M. Rogers, *Diffusion of Innovations*, 3rd edn. (Free Press, New York, 1983)
267. J. Roughgarden, R.M. May, S.A. Levin (eds.), *Perspectives in Ecological Theory* (Princeton University Press, Princeton, 2014)
268. Z. Ruan, G. Iñiguez, M. Karsai, J. Kertész, Kinetics of social contagion. Phys. Rev. Lett. **115**(21), 218702 (2015)
269. J. Ruths, D. Ruths, Control profiles of complex networks. Science **343**(6177), 1373–1376 (2014)
270. A.A. Saberi, Recent advances in percolation theory and its applications. Phys. Rep. **578**, 1–32 (2015)
271. M. Salehi, R. Sharma, M. Marzolla, M. Magnani, P. Siyari, D. Montesi, Spreading processes in multilayer networks. IEEE Trans. Netw. Sci. Eng. **2**(2), 65–83 (2015)
272. H. Sayama, I. Pestov, J. Schmidt, B. J. Bush, C. Wong, J. Yamanoi, T. Gross, Modeling complex systems with adaptive networks. Comput. Math. Appl. **65**(10), 1645–1664 (2013)
273. S.V. Scarpino, A. Allard, L. Hébert-Dufresne, Prudent behaviour accelerates disease transmission (2015). arXiv:1509.00801
274. I. Scholtes, N. Wider, R. Pfitzner, A. Garas, C.J. Tessone, F. Schweitzer, Causality-driven slow-down and speed-up of diffusion in non-Markovian temporal networks. Nat. Commun. **5**, 5024 (2014)
275. S.B. Seidman, Network structure and minimum degree. Soc. Networks **5**(3), 269–287 (1983)
276. F. Sélley, A. Besenyei, I.Z. Kiss, P.L. Simon, Dynamic control of modern, network-based epidemic models. SIAM J. Appl. Dyn. Syst. **14**(1), 168–187 (2015)
277. J.P. Sethna, *Statistical Mechanics: Entropy, Order Parameters and Complexity* (Oxford University Press, Oxford, 2006)
278. C.R. Shalizi, A.C. Thomas, Homophily and contagion are generically confounded in observational social network studies. Sociol. Methods Res. **40**(2), 211–239 (2011)
279. F. Shi, P.J. Mucha, R. Durrett, Multiopinion coevolving voter model with infinitely many phase transitions. Phys. Rev. E **88**(6), 062818 (2013)
280. I. Shmulevich, E.R. Dougherty, S. Kim, W. Zhang, Probabilistic Boolean networks: A rule-based uncertainty model for gene regulatory networks. Bioinformatics **18**(2), 261–274 (2002)
281. J.D. Skufca, E.M. Bollt, Communication and synchronization in disconnected networks with dynamic topology: Moving neighborhood networks. Math. Biosci. Eng. **1**(2), 347–359 (2004)
282. R. Smith? *Braaaiiinnnsss!: From Academics to Zombies* (University of Ottawa Press, Ottawa, 2011)

283. T.A.B. Snijders, The statistical evaluation of social network dynamics. Sociol. Methodol. **40**(1), 361–395 (2001)

284. T.A.B. Snijders, G.G. Van de Bunt, C.E.G. Steglich, Introduction to stochastic actor-based models for network dynamics. Soc. Networks **32**(1), 44–60 (2010)

285. A. Solé-Ribalta, M. De Domenico, N.E. Kouvaris, A. Díaz-Guilera, S.Gómez, and A. Arenas. Spectral properties of the Laplacian of multiplex networks. Phys. Rev. E **88**(3), 032807 (2013)

286. V. Sood, S. Redner, Voter model on heterogeneous graphs. Phys. Rev. Lett. **94**(17), 178701 (2005)

287. V. Sood, T. Antal, S. Redner, Voter models on heterogeneous networks. Phys. Rev. E **77**(4), 041121 (2008)

288. M. Starnini, A. Baronchelli, A. Barrat, R. Pastor-Satorras, Random walks on temporal networks. Phys. Rev. E **85**(5), 056115 (2012)

289. B. State, L. Adamic, The unequal adoption of equal signs (2013). Available at https://www.facebook.com/notes/facebook-data-science/the-unequal-adoption-of-equal-signs/10151927935438859

290. B. State, L. Adamic, The diffusion of support in an online social movement: Evidence from the adoption of equal-sign profile pictures, in *Proceedings of the 18th ACM Conference on Computer Supported Cooperative Work & Social Computing, CSCW '15* (ACM, New York, 2015), pp. 1741–1750

291. I. Stewart, M. Golubitsky, M. Pivato, Symmetry groupoids and patterns of synchrony in coupled cell networks. SIAM J. Appl. Dyn. Syst. **2**(4), 609–646 (2003)

292. S.H. Strogatz, *Nonlinear Dynamics and Chaos* (Addison-Wesley, Massachusetts, 1994)

293. S.H. Strogatz, From Kuramoto to Crawford: Exploring the onset of synchronization in populations of coupled oscillators. Physica D **143**(1–4), 1–20 (2000)

294. S.H. Strogatz, Exploring complex networks. Nature **410**(6825), 268–276 (2001)

295. M.P.H. Stumpf, M.A. Porter, Critical truths about power laws. Science **51**(6069), 665–666 (2012)

296. K. Suchecki, V.M. Eguíluz, M. San Miguel, Conservation laws for the voter model in complex networks. Europhys. Lett. **69**(2), 228 (2005)

297. J. Sun, E.M. Bollt, M.A. Porter, M.S. Dawkins, A mathematical model for the dynamics and synchronization of cows. Physica D **240**(19), 1497–1509 (2011)

298. J. Sun, E.M. Bollt, T. Nishikawa, Master stability functions for coupled nearly identical dynamical systems. Europhys. Lett. **85**(6), 60011 (2011)

299. T. Takaguchi, N. Masuda, P. Holme, Bursty communication patterns facilitate spreading in a threshold-based epidemic dynamics. PLoS ONE **8**(7), e68629 (2013)

300. H.G. Tanner, A. Jadbabaie, G.J. Pappas. Stable flocking of mobile agents, part i: Fixed topology, in *Proceedings of the 42nd IEEE Conference on Decision and Control, 2003*, vol. 2, pp. 2010–2015 (2003)

301. H.G. Tanner, A. Jadbabaie, G.J. Pappas, Stable flocking of mobile agents, part ii: Dynamic topology, in *Proceedings of the 42nd IEEE Conference on Decision and Control, 2003*, pp. 2016–2021 (2003)

302. T.J. Taylor, I.Z. Kiss, Interdependency and hierarchy of exact and approximate epidemic models on networks. J. Math. Biol. **69**(1), 183–211 (2014)

303. D. Taylor, F. Klimm, H.A. Harrington, M. Kramár, K. Mischaikow, M.A. Porter, P.J. Mucha, Topological data analysis of contagion maps for examining spreading processes on networks. Nat. Commun. **6**, 7723 (2015)

304. D. Taylor, S.A. Myers, A. Clauset, M.A. Porter, P.J. Mucha, Eigenvector-based centrality measures for temporal networks (2015). arXiv:1507.01266

305. M. Tizzoni, P. Bajardi, C. Poletto, J. Ramasco, D. Balcan, B. Gonçalves, N. Perra, V. Colizza, A. Vespignani, Real-time numerical forecast of global epidemic spreading: Case study of 2009 A/H1N1pdm. BMC Med. **10**(1), 165 (2012)

306. P. Trapman, On analytical approaches to epidemics on networks. Theor. Popul. Biol. **71**(2), 160–173 (2007)

307. A.L. Traud, P.J. Mucha, M.A. Porter, Social structure of Facebook networks. Physica A **391**(16), 4165–4180 (2012)

308. J. Ugander, L. Backstrom, C. Marlow, J. Kleinberg, Structural diversity in social contagion. Proc. Natl. Acad. Sci. U. S. A. **109**(16), 5962–5966 (2012)
309. E. Valdano, L. Ferreri, C. Poletto, V. Colizza, Analytical computation of the epidemic threshold on temporal networks. Phys. Rev.X **5**(2), 021005 (2015)
310. T.W. Valente, *Network Models of the Diffusion of Innovations* (Hampton Press, New York, 1995)
311. T.W. Valente, Social network thresholds in the diffusion of innovations. Soc. Networks **18**(1), 69–89 (1996)
312. P. Van Mieghem, Exact Markovian SIR and SIS epidemics on networks and an upper bound for the epidemic threshold (2014). arXiv:1402.1731
313. P. Van Mieghem, *Graph Spectra for Complex Networks* (Cambridge University Press, Cambridge, 2013)
314. P. Van Mieghem, R. Van de Bovenkamp, Non-Markovian infection spread dramatically alters the susceptible-infected-susceptible epidemic threshold in networks. Phys. Rev. Lett. **110**(10), 108701 (2013)
315. F. Vázquez, V.M. Eguíluz, Analytical solution of the voter model on uncorrelated networks. New J. Phys. **10**(6), 063011 (2008)
316. F. Vázquez, S. Redner, Ultimate fate of constrained voters. J. Phys. A Math. Gen. **37**(35), 8479–8494 (2004)
317. F. Vázquez, P.L. Krapivsky, S. Redner, Constrained opinion dynamics: Freezing and slow evolution. J. Phys. A: Math. Gen. **36**(3), L61–L68 (2003)
318. A. Vespignani, Modelling dynamical processes in complex socio-technical systems. Nat. Phys. **8**(1), 32–39 (2012)
319. C.L. Vestergaard, M. Génois, Temporal Gillespie algorithm: Fast simulation of contagion processes on time-varying networks. PLoS Comput. Biol. **11**(10), e1004579 (2015)
320. T. Vicsek, A. Zafeiris, Collective motion. Phys. Rep. **517**(3–4), 71–140 (2012)
321. T. Vicsek, E. Czirók, E. Ben-Jacob, I. Cohen, O. Shochet, Novel type of phase transition in a system of self-driven particles. Phys. Rev. Lett. **75**(6), 1226–1229 (1995)
322. E. Volz, L.A. Meyers, Susceptible–infected–recovered epidemics in dynamic contact networks. Proc. R. Soc. Lond. B Biol. Sci. **274**(1628), 2925–2934 (2007)
323. R.S. Wang, A. Saadatpour, R. Albert, Boolean modeling in systems biology: An overview of methodology and applications. Phys. Biol. **9**(5), 055001 (2012)
324. W. Wang, M. Tang, H.-F. Zhang, Y.-C. Lai, Dynamics of social contagions with memory of nonredundant information. Phys. Rev. E **92**(1), 012820 (2015)
325. Z. Wang, L. Wang, A. Szolnoki, M. Perc, Evolutionary games on multilayer networks: A colloquium. Eur. Phys. J. B **88**(5), 124 (2015)
326. L. Warnke, O. Riordan, Explosive percolation is continuous. Science **333**(6040), 322–324 (2011)
327. L. Warnke, O. Riordan, Achlioptas process phase transitions are continuous. Ann. Appl. Probab. **22**(4), 1450–1464 (2012)
328. D.J. Watts, A simple model of global cascades on random networks. Proc. Natl. Acad. Sci. U.S.A. **99**(9), 5766–5771 (2002)
329. L. Weng, A. Flammini, A. Vespignani, F. Menczer, Competition among memes in a world with limited attention. Sci. Rep. **2**, 335 (2012)
330. L. Weng, F. Menczer, Y.-Y. Ahn, Virality prediction and community structure in social networks. Sci. Rep. **3**, 2522 (2013)
331. G.B. Whitham, *Linear and Nonlinear Waves*. Pure and Applied Mathematics (Wiley-Interscience, New York, 1974)
332. D. Witthaut, M. Timme, Braess's paradox in oscillator networks, desynchronization and power outage. New J. Phys. **14**(8), 083036 (2012)

333. O. Yağan, V. Gligor, Analysis of complex contagions in random multiplex networks. Phys. Rev. E **86**(3), 036103 (2012)

334. Y. Zou, T. Pereira, M. Small, Z. Liu, J. Kurths, Basin of attraction determines hysteresis in explosive synchronization. Phys. Rev. Lett. **112**(11), 114102 (2014)

335. G. Zschaler, T. Gross, Largenet2: An object-oriented programming library for simulating large adaptive networks. Bioinformatics **29**(2), 277–278 (2013)

Printed in the United States
By Bookmasters